"The climate emergency mandates that cities transition to low carbon. Dr Nochta shows how the complexity of urban networks can be better understood and aid the practice of transition management. Her novel and accessible interdisciplinary approach is essential reading for those working in and studying energy policy, urban governance and change management."

Prof. Chris Skelcher, *Professor Emeritus Institute of Local Government Studies (INLOGOV), University of Birmingham, UK*

"This book provides valuable insights into the structures and processes involved in urban governance, and the response of local authorities to the climate emergency. Based on in-depth case-studies, it is essential reading to appreciate the role that cities can play in the transition to decarbonised societies."

Dr Jonathan Radcliffe, *Reader in Energy Systems and Policy School of Chemical Engineering, University of Birmingham, UK*

"This volume uses network and governance perspectives to provide a rich account of myriad contextual factors affecting low carbon transitions. The rich case studies provide a vivid account of the terrain, and highlight key issues, making a valuable contribution for anyone wishing to better study, understand, or influence urban transitions."

Dr Dai Morgan, *Deputy Course Director, MPhil in Engineering for Sustainable Development Department of Engineering, University of Cambridge, UK*

"This book is a must read for anyone interested in how we can transition towards a low carbon society. Its emphasis on the importance of governance networks, and the influence of the local contexts in which they sit, is unique and an essential contribution to the field."

Dr Louise Reardon, *Lecturer in Governance and Public Policy Institute of Local Government Studies (INLOGOV), University of Birmingham, UK*

"This book is an important step in furthering our understanding of network forms of governance in cities. The topic is approached in a clear, systematic way building on existing research, but challenging and testing theories from a basis of local interpretation and socio-economic development. Planners, engineers and policy makers will find new evidence and a nascent approach to successful low-carbon transitions."

Mr Peter Braithwaite, *Director of Engineering Sustainability*
Birmingham Centre for Resilience Research and Education,
University of Birmingham, UK

Network Governance and Energy Transitions in European Cities

This book investigates and evaluates the opportunities and limitations of network governance in building local capacity for energy infrastructure governance.

Presenting a comparative analysis of three city cases from across Europe – Birmingham, Frankfurt and Budapest – this book demonstrates how local factors shape the prospect of network governance to support low-carbon energy transitions. It maps out existing governance networks, highlighting the actors involved and their interactions with one another, and also discusses the role and embeddedness of networks in the urban governance of low-carbon energy. Drawing on case study evidence, Nochta develops a comparative analysis which discusses the intricate connections between network characteristics, context and impact. It highlights that organisational fragmentation: the complexity of the low-carbon energy problem and historical developments all influence network characteristics in terms of the degree of integration and vertical (hierarchical) power relationships among network actors. Overall, the book concludes that understanding such links between context and networks is crucial when designing and implementing new governance models aimed at facilitating and governing low-carbon urban development.

Low-Carbon Energy Transitions in European Cities will be of great interest to scholars of energy policy, urban governance and sustainability transitions.

Timea Nochta is a Research Associate at the Centre for Smart Infrastructure and Construction, University of Cambridge. She completed her doctoral studies at the Institute of Local Government Studies, University of Birmingham, United Kingdom.

Routledge Focus on Energy Studies

Limits to Terrestrial Extraction
Edited by Robert E. Kirsch

Network Governance and Energy Transitions in European Cities
Timea Nochta

For more information about this series, please visit: https://www.routledge.com/Routledge-Focus-on-Energy-Studies/book-series/RFENS

Network Governance and Energy Transitions in European Cities

Timea Nochta

Routledge
Taylor & Francis Group
LONDON AND NEW YORK

First published 2021
by Routledge
2 Park Square, Milton Park, Abingdon, Oxon OX14 4RN

and by Routledge
52 Vanderbilt Avenue, New York, NY 10017

Routledge is an imprint of the Taylor & Francis Group, an informa business

British Library Cataloguing in Publication Data
A catalogue record for this book is available from the British Library

Library of Congress Cataloging-in-Publication Data
A catalog record has been requested for this book

ISBN: 978-0-367-46506-3 (hbk)
ISBN: 978-1-003-02916-8 (ebk)

Typeset in Times New Roman
by Taylor & Francis Books

To Magdolna and Mihály Nochta

Contents

Illustrations

Figures

Tables

Acknowledgements

This book is based on the author's doctoral research conducted at the University of Birmingham. Funding for this research was provided by Climate KIC (Knowledge & Innovation Community), part of the European Institute of Innovation and Technology (EIT), a body of the European Union. The study was affiliated with Climate KIC's Transition Cities programme.

Introduction

Today we are living in an era of *complexity.* Issues such as climate change and environmental protection, economic globalisation, global migration, or genome editing facilitate polarisation and heated ethical and political debates worldwide. Emphasising the controversies intrinsic to contemporary societies, such issues highlight misalignments between the inherited systems of governance (political and economic), the increasingly networked nature of societies, and the systemic character of the problems they are facing. In an attempt to find solutions to such problems, public policy and administration scholars and practitioners have been proposing more collaborative, networked modes of governance in recent decades. However, there remains a lack of understanding and evidence around the real-world contributions of networked forms of governance to solving contemporary complex problems (cf. Lewis, 2011). To help close this gap, this book seeks answers to the questions whether, and how, network governance can contribute to transitions towards more sustainable urban futures in different places with diverse contextual characteristics.

This complexity argument is often used in public policy-making and implementation to explain the emergence of the 'network governance' concept (Klijn and Koppenjan, 2015). Network governance, the argument goes, is a response to the failure of the traditional hierarchical and the competitive market models of social coordination in coping better with contemporary complex problems. Such problems, for example sustainable urban development, cannot be *solved* as they involve a multiplicity of societal actors with diverse (and often contradicting) interests and perceptions of the problem itself, the potential solutions and the actions to be taken. In addition, they do not fit neatly into any existing systems of social coordination such as national jurisdictions, organisations or sectors of the economy.

Dealing with complex problems thus requires developing interaction processes – information exchange, negotiations, collaboration, bargaining, etc. – among the affected and/or interested societal actors. Such interactions, over time, may contribute to developing a deeper understanding of the problems in question, the acceptable solutions and courses of action in a more inclusive and participatory way. The concept of 'network governance' aims to capture the importance and influence of these interaction patterns on outcomes of policy decision-making and implementation/delivery. It refers to a mode of social coordination that is distinctly different both from hierarchical command-and-control mechanisms and free market competition. In contrast, the term 'governance network' is used to denote the set of interacting actors. This interpretation acknowledges that interactions among network actors are not exclusively based on network governance as the dominant coordination mechanism.

This distinction is important: network governance, in practice, takes place alongside market and hierarchical coordination mechanisms and structures. Whilst much of the literature often suggests a linear development path along which clear transformations from hierarchies to markets (from the 1970s) and latterly to networks took place, empirical evidence gathered through in-depth case studies paints a rather different picture (Skelcher, Sullivan and Jeffares, 2013). To a varying extent, in place-specific and often temporary combinations, all three approaches remain in use in different policy domains (Considine and Lewis, 2003; Lynn, 2011). This in turn highlights the importance of considering the historical socio-political developments, cultures and locally relevant constraints and enablers that maintain distinct path dependencies – i.e. the 'context' – in discussing the role of network governance in facilitating and governing low-carbon development in different places.

To unravel the implications of context on networks and impact, the study presented in this book investigates transitions towards low-carbon urban futures in different places, with a focus on the urban scale. The city cases cover a broad geographic area and diverse national contexts within Europe. The distinct historical development trajectories of Birmingham (United Kingdom), Frankfurt-am-Main (Germany), and Budapest (Hungary) enable the showcasing and analysis of differences and similarities among local governance networks, and their contribution to facilitating and governing the decarbonisation of city systems.

Most of the currently available governance frameworks and models aiming to enable low-carbon urban development rely in some form on network governance. The Transition Management (TM) model and the intermediation perspective explicitly refer to concepts of networks and network governance. In TM, the goal is to organise new networks that can develop into 'advocacy coalitions' for advancing local sustainability transitions (Frantzeskaki, Loorbach and Meadowcroft, 2012). In contrast, the intermediation perspective emphasises existing networks and aims to build capacity for sustainability transitions through reducing societal and organisational fragmentation via intermediaries undertaking bridging and brokerage roles (Hodson and Marvin, 2010). Consideration for the impact of the local 'context' (made up of historical socio-political developments, cultures and locally relevant constraints and enablers that maintain distinct path dependencies) on network structures and processes, and by extension the potential of governance networks to support low-carbon transition is, however, often lacking. This is problematic, because this lack of engagement with the conditions in which networks operate may lead to potentially overestimating their potential to contribute to sustainability transitions (Lewis, 2011; Nochta and Skelcher, 2020).

Focus: Low-carbon energy transitions in European cities

> In what was widely denounced as one of the worst outcomes in a quarter-century of climate negotiations, United Nations talks ended early Sunday morning with the United States and other big polluters blocking even a nonbinding measure that would have encouraged countries to adopt more ambitious targets for reducing greenhouse gas emissions next year.
>
> (Sengupta, 2019)

The most recent climate negotiations, taking place on the 25th Conference of Parties (COPs) of the United Nations Framework Convention on Climate Change (UNFCCC) in Madrid, have widely been considered a failure and 'lost opportunity' (Cleetus, 2019). This failure comes at a time when it is becoming increasingly clear that keeping global warming below the limits set out by the Paris Agreement[1] will require significant social, economic and technological changes in the upcoming decades, rapid decarbonisation and effectively reaching a carbon-neutral global economy by the middle of the 21st century.

Achieving this goal necessitates concentrated action at the global, national and local political-organisational levels, as well as developing appropriate and effective multilevel governance systems. The role and importance of the local level (cities and regions) in tackling the climate crisis has been acknowledged in the Paris Agreement.

The opportunities identified include scaling up local (urban or neighbourhood) initiatives, maintaining and promoting cooperation between local level actors to mitigate emissions, and building resilience and reducing vulnerability by adapting to the adverse effects of climate change (UNFCCC, 2015).

Alongside international negotiations among nation states, and in response to the failures, cities and regions started taking a more proactive role and decisive position on carbon emissions reduction (Bulkeley, Broto and Edwards, 2015; Bai et al., 2018). Whilst early initiatives were concentrated in a small number of pioneer cities mostly from Europe and North America, today most cities and towns have sustainable development plans, and carbon emissions reduction goals (Broto and Bulkeley, 2013). However, oversight, key powers and the resources necessary to deal with complexities involved in the sustainable reorganisation of critical urban social and technical systems (for example, infrastructures and services that depend on them) are often lacking at the local level (Khan, 2013).

The decarbonisation of local energy systems is a crucial part of sustainable urban development. However, the governance of energy supply is typically nationally organised in most countries. Despite varying levels of authority and capacity for energy systems operation and management at the urban scale, energy policy-making and regulatory tasks are, in most cases, undertaken at higher organisational levels (i.e. European and national in Europe). Decentralised technologies associated with more sustainable local energy systems (such as heat and power co-production, district heat and steam networks, waste-to-energy schemes, and solar photovoltaic and thermal energy), and improving energy efficiency in buildings, industrial processes and transport, have opened up opportunities on the local level to influence the trajectory and pace of decarbonisation in the energy sector. These emerging solutions can be linked to services typically offered by Local Authorities, including waste and sewage management, education and social housing, and to social agendas, such as reducing fuel poverty and joblessness.

This situation creates an opportunity (and need) to re-evaluate the potential and role of city authorities in energy systems governance. The associated systemic reorganisation of infrastructures and their governance is nevertheless far from being straightforward. While governance networks may offer a solution to strengthen local governance capacity *in theory*, their impact *in practice* depends on a variety of factors. The study presented in this book aims to uncover what these factors are, and how they influence local governance networks. Beyond the structural characteristics of the governance of energy systems, it argues that socio-political, cultural and historical factors are also part of the *context* in which sustainable energy transitions need to take place, and in which governance networks need to operate.

Key contributions: Networks, context and impact

Contributions are derived through the comparative analysis of the governance networks of low-carbon energy in the cities of Birmingham, Frankfurt and Budapest. The analysis highlights both potential benefits as well as shortcomings of network forms of governance in different urban settings. Networks are often praised for

> creat[ing] space for interaction and multi sectoral co-operation between various organizations; for facilitating informed decision-making based on knowledge exchange and deliberation between interdependent stakeholders; and for building engagement to achieve the negotiated goals.
>
> (Nochta and Skelcher, 2020)

However, they can also contribute to maintaining the status quo because they 'act as conduits to agent power' and network structures and processes 'reflect the power of agents' (Christopoulos, 2017, p. 57). As such, networks may also facilitate increased private sector influence in public policy-making and implementation. This book argues that these are important considerations which are largely overlooked in the literature on urban sustainability transitions which prescribe networks as the *solution* to the failures of current societal systems to facilitate low-carbon development.

In order to investigate this issue, a novel framework is developed which emphasises the importance of considering contextual factors related to historical socio-political developments, cultures and locally relevant constraints and enablers that maintain distinct path dependencies in different cities. Analysing the structural and procedural features of potentially already existing networks in different cities, whose remit is relevant to local low-carbon energy, and linking them to contextual factors, can contribute to developing a more nuanced understanding of the potential of network forms of governance to support low-carbon development in different urban settings. By doing so, it can also help in evaluating the appropriateness of different models for governing transitions.

The key contributions of the book therefore regard first a conceptual framework which links local contexts, its impact on existing networks, and networks' potential to deliver on local low-carbon goals. Second, a comparative research design based on mixed methods approach to data collection and analysis combining quantitative network and qualitative discourse analyses for the study of multi-actor governance structures, processes and outcomes. Third, recommendations and insights for identifying niche roles that network forms of governance can play in different cities.

The structure of the book

This book has six chapters. Chapter 1 discusses the concepts of networks and network governance in more detail, and presents how they appear in the literature on urban sustainability transitions. Based on this discussion, the chapter sets out a critically informed theoretical and conceptual framework for investigating the role of networks and network governance in low-carbon transitions in different cities. Chapter 2 presents the methodology and the comparative analysis of the existing energy transition networks in Birmingham, Frankfurt and Budapest. The organisational network analysis is used to highlight similarities and differences among the three local networks of low-carbon energy. It provides a high-level overview of the existing networks and the actors involved, as well as initial insights on the context-dependence of network forms of governance.

The next three chapters (3, 4 and 5) focus on describing the local contexts in which the networks have emerged and operate in the cities of Birmingham (United Kingdom), Frankfurt-am-Main (Germany) and Budapest (Hungary). The qualitative analysis presented in these chapters thus further investigates, complements and validates the results of the network analysis presented in Chapter 2. The in-depth case studies link change factors associated with the emergence of governance networks to impact in terms of shaping the trajectories of energy transitions and transition networks in the different cities. These describe, first, the dispersion of power, authority and resources in energy systems governance (societal and organisational fragmentation); second, the locally perceived complexity of energy transitions (complexity in contemporary decision-making); third, the development of the sustainability agenda over time (new societal dynamics related to a shift from 'government' to 'governance').

On the basis of the individual case studies, Chapter 6 presents the conclusions of the comparative and case study analyses. It concludes that contextual change factors influence the structure of the energy transitions networks in terms of degree of integration and vertical (hierarchical) power relationships among network actors. This supports the argument that building a more nuanced understanding of the links between context and networks is crucial when designing and implementing new governance initiatives aimed at facilitating and governing low-carbon urban transitions in different cities.

Applying TM or intermediation perspective informed strategies to facilitate low-carbon development is an intervention into existing webs of relationships and interactions rather than *de novo* institution-building. As such, 'contextualising' energy transition networks can also provide insights on how to improve networks' impact in terms of contribution to facilitating and governing energy transitions. Practical recommendations regard options for developing more targeted interventions which take into account contextual enabling and constraining factors, existing network relationships between locally relevant actors, as well as Local Authorities' capacity and capability to coordinate transition processes.

Note

1 The Paris Agreement is an agreement of the Parties to the United Nations Framework Convention on Climate Change (UNFCCC) which aims at accelerating and intensifying the actions and investments needed for a sustainable low-carbon future, signed in 2016. It represents a commitment of keeping global temperature rise this century well below 2 degrees Celsius above pre-industrial levels (pursuing efforts to limit the increase even further to 1.5 degrees Celsius). Source: unfccc.int/process-and-meetings/the-paris-agreement/what-is-the-paris-agreement

References

Bai, X.*et al.* (2018) 'Six research priorities for cities and climate change', *Nature*, pp. 23–25. doi:10.1038/d41586-018-02409-z.

Broto, V. C. and Bulkeley, H. (2013) 'A survey of urban climate change experiments in 100 cities', *Global Environmental Change*, 23(1), pp. 92–102. doi:10.1016/j.gloenvcha.2012.07.005.

Bulkeley, H., Broto, V. C. and Edwards, G. A. S. (2015) *An Urban Politics of Climate Change: Experimentation and the Governing of Socio-Technical Transitions.* Routledge. doi:10.1017/CBO9781107415324.004.

Christopoulos, D. (2017) 'Governance networks in politics', in B. Hollstein, W. Matiaske and K.-U. Schnapp (eds), *Networked Governance: New Research Perspectives*, pp. 55–70. Springer. doi:10.1007/978-3-319-50386-8_4.

Cleetus, R. (2019) Reflections from COP25 in Madrid – Union of Concerned Scientists. Available at: https://blog.ucsusa.org/rachel-cleetus/reflections-from-cop25 (Accessed: 28 February 2020).

Considine, M. and Lewis, J. M. (2003) 'Bureaucracy, network, or enterprise? Comparing models of governance in Australia, Britain, the Netherlands, and New Zealand', *Public Administration Review*, 63(2), pp. 131–140. doi:10.1111/1540-6210.00274.

Frantzeskaki, N., Loorbach, D. and Meadowcroft, J. (2012) 'Governing societal transitions to sustainability', *International Journal of Sustainable Development*, 15(1–2), pp. 19–36. doi:10.1504/IJSD.2012.044032.

Hodson, M. and Marvin, S. (2010) 'Can cities shape socio-technical transitions and how would we know if they were?', *Research Policy*, 39(4), pp. 477–485. doi:10.1016/j.respol.2010.01.020.

Khan, J. (2013) 'What role for network governance in urban low carbon transitions?', *Journal of Cleaner Production*, 50, pp. 133–139. doi:10.1016/j.jclepro.2012.11.045.

Klijn, E. H. and Koppenjan, J. (2015) *Governance Networks in the Public Sector.* Routledge. doi:10.4324/9781315887098.

Lewis, J. M. (2011) 'The future of network governance research: Strength in diversity and synthesis', *Public Administration*, 89(4), pp. 1221–1234. doi:10.1111/j.1467-9299.2010.01876.x.

Lynn, L. E. (2011) 'The persistence of hierarchy', in M. Bevir (ed.), *The SAGE Handbook of Governance*, pp. 218–236. SAGE.

Nochta, T. and Skelcher, C. (2020) 'Network governance in low-carbon energy transitions in European cities: A comparative analysis', *Energy Policy*, 138. doi:10.1016/j.enpol.2020.111298.

Sengupta, S. (2019) 'U.N. climate talks end with few commitments and a "lost" opportunity', *The New York Times.* Available at: www.nytimes.com/2019/12/15/climate/cop25-un-climate-talks-madrid.html (Accessed: 28 February 2020).

Skelcher, C., Sullivan, H. and Jeffares, S. (2013) *Hybrid Governance in European Cities: Neighbourhood, Migration and Democracy.* Springer.

UNFCCC (2015) The Paris Agreement. Available at: https://unfccc.int/process-and-meetings/the-paris-agreement/the-paris-agreement (Accessed: 26 February 2020).

1 Network governance in urban sustainability transitions

Introduction

This chapter presents the development of the concepts of networks and governance in public policy-making and implementation, and links the different perspectives emerging from this field to the use of network governance in theories and frameworks for governing urban sustainability transitions. The term 'governance' is used to refer to the outcome of all coordination mechanisms developing from hierarchical, market-style and network interactions among multiple actors throughout this book (Meuleman, 2008; Torfing, 2012). In contrast, 'network governance' is understood as a process that emerges from interactions (e.g. collaboration, negotiation or bargaining) among network actors as a horizontal mode of coordination (Rhodes, 1997; Klijn and Skelcher, 2007). The 'network' (governance network) denotes the set of (public or private) actors connected through the web of interactions (Torfing, 2005).

The importance of facilitating interactions and network forms of governance among societal actors as a form of coordination is often emphasised in studies on urban sustainability transitions. Here, two specific approaches are discussed in detail which explicitly refer to networks in governing low-carbon transitions: Transition Management (Loorbach, 2010; Nevens et al., 2013; Wittmayer et al., 2018) and the intermediation perspective (Hodson, Marvin and Bulkeley, 2013; Kivimaa et al., 2019). However, the role of local contextual factors in terms of their impact on network structures and processes, and by extension the potential of governance networks to support low-carbon transitions, remains unaddressed. This leads to a risk of potentially overestimating the impact that governance networks can deliver in different cities, without sufficiently engaging with the conditions necessary for them to function *well* (Lewis, 2011; Nochta and Skelcher, 2020).

In order to unpack this issue, the chapter reviews the emergence of network forms of governance in academic research and practice from a historical perspective. It presents the different ways in which network concepts have appeared in public policy-making and implementation, such as policy networks, inter-organisational (service delivery) and collaborative inter-governmental networks. Building a more nuanced understanding of networks and governance is important to highlight gaps in the conceptualisation of networks and the network mode of governance in the literature on urban sustainability transitions. The identified gaps form the basis of the research agenda set out to be addressed through the comparative analysis of the case studies of low-carbon transition networks in three European cities.

The emergence of the network concept – a historical overview

The shift from 'government' to 'governance' in the field of public policy and administration refers to a move away from the state-centric hierarchical model towards a more polycentric social-organisational context in which a multiplicity of state and non-state actors participate in the formulation and implementation of public policy (Mayntz, 2019). Historically, the hierarchical model of public policy-making and implementation, based on the centrality of the state, was predominant in the post-war era Europe and North America. Here, the state was conceptualised as a sovereign entity, governing through command-and-control mechanisms (Bevir and Rhodes, 2016). Starting from the 1970s, parallel developments including the advancement of social science research, technological progress and periodic economic crises contributed to a shift in thinking about the role of the state in society in conjunction with the emergence of the view that the welfare state had become overloaded, unaffordable and ineffective (Bevir, 2011). The task set out, therefore, was to 'modernise' the public sector through the adoption of techniques borrowed from the market to enhance the effectiveness, efficiency and responsiveness of the state (Pollitt and Bouckaert, 2011).

Market-style reforms were aimed at focusing state operations on core tasks and responsibilities, and introducing a managerial approach to public administration. Seeking improvements in efficiency and effectiveness, these 'managerial reforms' (or 'New Public Management'; Dunleavy and Hood, 1994) included the internal restructuring of the public sector through separating politics (strategy) and administration (execution), and dividing large organisations into small, specialised units. In terms of process, public service provision was to be improved through introducing performance-oriented evaluation (output measurement and key performance indicators) and creating individualised incentives for public managers (Pollitt and Bouckaert, 2011). Regarding the role of the public sector vis-à-vis the private domain, the routine contracting-out of service provision outside the identified core services contributed to a blurring of boundaries between the public and private spheres (Homburg, Pollitt and van Thiel, 2007). The privatisation of state assets created new revenue flows and boosted the public budget. The role of the public sector has been reoriented towards a steering role of safeguarding free and fair market competition and consumer rights. Citizens in these new arrangements have been redefined as 'service users' or 'customers'. Although the reforms have initially concentrated on the Anglo-Saxon world from where they had originated, over time and in diverse forms they spread to other countries, creating location-specific combinations with different cultures and structures of governing (Pollitt, 2003).

It took a few decades for the shortcomings and limitations of the reforms to become more apparent. Managerial reforms are still being implemented in different places and policy domains today. In recent years, however, a variety of critical debates emerged centred on the unintended consequences such as decreased coordination capacity in the fragmented organisational landscape (Skelcher, 2000), and shortcomings in terms of accountability and the democratic quality in policy-making (Kersbergen and Waarden, 2004; Drechsler, 2005). Coupled with the effects of globalisation (i.e. increased collaboration and exchange between nation states), heightened awareness of the complexities involved in decision-making processes, and the changing societal perceptions about the role of public and private actors in public policy-making and implementation, these problems once again opened the way towards a quest for new forms of societal organising (Torfing, 2005).

In response, concepts of governance, networks, partnerships and participation started emerging in key debates around public policy and implementation from around the millennium (Pollitt and Bouckaert, 2011). Termed collectively as 'New Public Governance', this wave of ideas is 'predicated upon the existence of a plural state and a pluralist state and seeks to understand the development and implementation of public policy in this context' (Osborne, 2010, p. 7). These developments led Rhodes (1996) to conclude that hierarchies (based on command-and-control mechanisms), markets (based on free competition) and networks (based on interactions such as deliberation, negotiation, collaboration or bargaining) constitute three ideal types of governing modes, from which governments can choose according to the particular task or problem at hand. However, the extent to which these three different approaches can be deployed independently of one another *in practice*, and the impact this has on the outcomes of real-world decision-making processes, remains unclear.

Different perspectives on the role of networks in governance

A large body of academic and practice-oriented literature exists which investigates issues of the dispersion of power, authority and resources, and the related processes of participation, collaboration, conflict and coordination. Klijn and Koppenjan (2015) argue that the literature that is relevant to network forms of governance is concentrated around three central foci of policy networks, inter-organisational delivery networks and collaborative/inter-governmental governance (cf. Bevir and Richards, 2009). These different perspectives have their origins in different research fields – political science, organisational science and public administration respectively.

First, the policy network perspective investigates networks of actors involved in, and their influence on, particular policies or policy-making in specific domains/sectors, problematising access, mobilisation, power and influence (Börzel, 1998). It therefore emphasises differentiation between more or less powerful actors, and vertical links and flows of influence among them, rather than horizontal collaboration. Consequently, policy networks do not necessarily result in network governance as a mode of horizontal coordination (Lewis, 2011, p. 1222). This observation is particularly relevant to the sustainability domain in which networks may refer to interest groups that gather around key government departments and exert their influence over climate change politics. Networks are thus seen as vehicles for dominant actors – who most often have an economic interest in the fossil-fuel industry in some shape or form – to exert their influence over policy in an attempt to block outcomes that may have a negative impact on the economy and their profits (see for example Bulkeley, 2000).

Second, the inter-organisational (service delivery) perspective has been most prevalent in the United States, focusing mainly on networks in policy implementation and integrated service delivery (Provan and Milward, 2001). Here, network interactions are seen as being facilitated by interdependencies in terms of resources, responsibilities and tasks in a fragmented organisational context. They may also be mandated and/or overseen by higher-level governmental bodies (Provan and Kenis, 2008). The structural characteristics of the network are important in conjunction with their role in effective and efficient implementation and delivery.

Related to the sustainability domain, this type of interpretation of the network governance concept has been used extensively to study inter-organisational emergency management networks in which relationships among different organisations are produced and maintained by the particular need for coordinated responses to natural disasters (Kapucu et al., 2010; Kapucu and Demiroz, 2017).

Third, the empirical focus of the research on collaborative / inter-governmental governance networks is the emergence of collaborative decision-making processes in situations where no functioning hierarchical order exists, i.e. in spaces characterised as 'institutional voids' (Hajer, 2003). The weak and/or absent hierarchies allow for horizontal relationships and collaboration-type interactions to emerge among network participants (Scharpf, 2001). In the sustainability domain, the management of environmental resources which cross national boundaries has become a key focus for research on collaborative inter-governmental governance. Here, network settings give rise to processes of information-sharing, capacity-building and rule-setting in order to achieve mutually beneficial outcomes (for an example see Gerlak, 2015).

Whilst acknowledging that, in certain domains, policy-making, implementation and service delivery are not – and cannot be – controlled by the state as a sovereign, monolithic entity, the different perspectives also highlight the diversity of processes that may arise in network settings. Moreover, although achieving certain goals or completing tasks may necessitate interactions among various actors, these interaction patterns allow for both flows of information and flows of influence (Lewis, Baeza and Alexander, 2008). The network concept can be operationalised either as a metaphor (as in the policy network perspective) or as a useful analytical framework (inter-organisational networks). Networks may be self-organising and emerge organically (inter-governmental networks) or mandated and coordinated by higher levels of government (inter-organisational networks).

More recently, some scholars have been arguing that, despite the differences, the various perspectives started converging and informing one another (Klijn and Koppenjan, 2012). One consequence of this turn is that 'network governance' in this body of work has been redefined as a mode of coordination in which network interactions have an impact, although not deterministic, on outcomes in policy-making and implementation – in addition to hierarchical and market-logic driven relationships – also leading to the use of the terms 'governance' and 'network governance' interchangeably. In contrast, others make the point that the different approaches consider distinct empirical cases (different types of networks), focus on interactions of different natures (influence, coordination, collaboration) and different sets of problems. Therefore, rather than aiming for convergence, the most appropriate analytical framework should be applied depending on the case and questions at hand (Lewis, 2011). Discussion on the diversity of networks and network processes, and their implications for decision-making, is, however, limited in the literature on urban sustainability transitions.

Networks in urban low-carbon transitions

Governance models for low-carbon transitions have traditionally been focusing on the national scale where relevant powers and authority are concentrated (Loorbach, 2010). Historically, the dominance of nation states has also been a feature of international negotiations for climate change mitigation and carbon emissions reduction. In the past few decades, however, cities have continued to gain importance as key sites for action to tackle climate change. Local and regional governments have been represented on the Conferences of Parties (COPs) of the United Nations Framework Convention on Climate Change (UNFCCC) through the Local Governments and Municipal Authorities (LGMA) Constituency since 1995 (LGMA, no date). More recently, the Paris Agreement developed at COP21 in 2015, recognised the need for action on multiple levels of government, including cities and regions, in addressing climate change (UNFCCC, 2015). Since 2015, city and region representatives have been calling for a formal role for local and regional governments in global climate governance and in the implementation of the Paris Agreement, complementing national pledges (European Committee of the Regions, 2018).

Whilst early urban climate change mitigation initiatives were confined to a number of pioneering cities in Europe and North America, interest in, and appetite for, taking action locally has been growing in tandem with the increasingly important role of the sub-national level in international negotiations. Most cities today engage with the issue of sustainable urban development, and as part of the local sustainability agenda, have carbon emissions reduction ambitions and targets (Broto and Bulkeley, 2013). Urban sustainability and low-carbon transitions are considered in various strands of urban studies literatures. Sustainable urban development has often been discussed in the context of rapid urbanisation and its environmental consequences (Mcgranahan and Satterthwaite, 2014). A large body of work developed investigating the diverse ways in which efforts to respond to climate change materialise in various cities globally (Bulkeley and Betsill, 2005; Bulkeley and Broto, 2013). A socio-technical perspective has been developed to analyse and support the low-Carbon transitions in urban infrastructure systems (Hodson and Marvin, 2010; Nevens et al., 2013).

These different, although overlapping, strands of research present climate change mitigation and adaptation, sustainable development and low-carbon innovation in the urban context as issues characterised by high levels of complexity and uncertainty. They highlight that constructing a universally accepted view on the nature of the problem to be solved, possible solution options and the course of action to be taken, is inherently problematic.

The arguments that underpin this assertion include the following. There are significant knowledge gaps in accurately estimating the local impacts of global climate change in different geographical locations (McCarthy, Best and Betts, 2010). This in turn results in difficulties with balancing mitigation and adaptation strategies and understanding the associated trade-offs (Anguelovski and Carmin, 2011). Currently functioning urban infrastructure systems, on which millions depend, require large-scale restructuring in conjunction with decarbonisation – at the same time they exhibit high levels of resistance to change (Bulkeley and Broto, 2013; Bulkeley, Broto and Edwards, 2015). In the contemporary globalised marketplace, cities and regions need to compete for investment, and the pursuit of economic growth often overrides environmental concerns (Bai, 2007). In addition, the relevant powers, authority and responsibilities that are necessary to formulate and implement policies and initiatives are dispersed among a multiplicity of societal actors operating at and between a variety of geographical and organisational scales in multilevel governance settings (Bulkeley and Betsill, 2013).

These factors result in a favourable environment for developing and experimenting with new governance models and frameworks to tackle the challenge of urban low-carbon transformations. Many of these, at least implicitly, incorporate ideas reminiscent of networks and network governance. One of the most influential frameworks, Transition Management (TM) explicitly aims at governing and managing urban low-carbon transitions through networks (Loorbach, 2010). In TM, low-carbon transitions denote 'fundamental transformation[s] towards more sustainable modes of production and consumption' (Markard, Raven and Truffer, 2012, p. 955). Transitions thus involve changes in 'markets, user practices, infrastructures, cultural discourses, policies and governing institutions' (Nevens et al., 2013, p. 112) to support technological innovation, i.e. the roll-out of carbon-neutral technologies.

TM's conceptualisation of governance networks follows from an emphasis on the complexity of facilitating the decarbonisation of city systems. As complex problems, low-carbon transitions necessarily involve a variety of interested and/or affected actors from the public, private and third sectors. 'Managing' complexity in turn requires a development of information exchange, resource pooling and coordination mechanisms to develop problem definitions, acceptable solution options and commitment for implementation/delivery. In TM, this is envisaged to be achieved through setting up participative 'transition arenas' which provide space for interaction among interested and/or affected parties. Interactions in transition arenas are expected to lead to a shift towards 'more innovative forms of governance that build on multi-actor interaction' (Hölscher et al., 2019, p. 145) – in other words a more predominant role of horizontal network-type coordination in urban governance. As such transition arenas are themselves governance experiments, contributing not only to technological (decarbonisation) but also social-political (participative governance) change.

Initially, TM's focus was mainly on sectoral transitions (e.g. of infrastructures) at the national scale (Kern and Smith, 2008). Transition arenas mostly comprised professionals relevant to the sector and theme in question. Applying the TM model to cities therefore represents a shift towards a more geographically defined focus, with different types of actors available to act as arena participants, including social entrepreneurs or 'ordinary' citizens as well as professionals (Frantzeskaki, Wittmayer and Loorbach, 2014). This shift in focus led to more emphasis on bottom-up processes, empowerment and co-creation. For example, one variation of the TM approach applied to the urban context is the concept of Urban Transition Labs (Nevens et al., 2013). The Urban Transition Lab framework builds on the 'Urban Living Lab' idea to translate the TM approach to the urban setting. Urban Living Labs are real-world testbeds aimed at facilitating faster feedback loops between users and producers of innovation (Voytenko et al., 2016). Similarly, the Urban Transition Lab model stresses the importance of innovation and co-creation to develop visions and pathways towards low-carbon futures and facilitate implementation. Transition Labs are conceptualised as 'facilitated sites [by the transition manager(s)] for creating (social) innovation and within which social change agents can initiate or inflict urban sustainability transitions' (Nevens et al., 2013, p. 115).

Transition arenas, as well as Urban Transition Labs, are set up by transition manager organisations which are also responsible for steering the emerging governance processes – towards horizontal coordination (e.g. through 'empowerment' (Avelino, 2009) and selective participation (Grin, Rotmans and Loorbach, 2010)) and sustainable/low-carbon outcomes. The models therefore contain the assumption that an organisation exists within the city which has the powers and resources necessary to undertake the role of the transition manager. From a network perspective, the TM and Urban Transition Labs models aim to leverage the potential of horizontal relationships which give way to collaborative processes, and knowledge sharing and deliberation among a variety of societal actors whose views need to be considered in urban low-carbon transitions. There is some acknowledgement that interactions among certain actors within the city are also governed by hierarchical or market mechanisms. Transition managers therefore need to ensure that arena memberships are constructed in such way that maximises the potential for horizontal collaborative relationships to develop.

This is, however, far from being straightforward. Consequently, TM has often been critiqued for paying insufficient attention to issues of politics and power (Smith, Stirling and Berkhout, 2005). Adding to this, the issue of scale becomes important when TM is 'rescaled' from the national to the local level. As an example, the number of available key stakeholders is usually fairly limited in a city. It is not far-fetched to predict that there may be existing or historical relationships between these actors which can also influence interactions within newly established transition arenas. In addition, the role of non-local actors in urban transition arenas, associated cross-scale linkages and their impact on local transitions, are also seldom discussed.

In contrast, studies emphasising the role of 'intermediation' and 'intermediaries' to some extent acknowledge such existing relationships across the governance landscape relevant to the decarbonisation of urban infrastructure systems (Hodson and Marvin, 2010; Kivimaa et al., 2019). Intermediaries – the actors performing intermediation – can take diverse forms (e.g. individuals, public sector agencies, public–private partnerships, NGOs), and operate within and between different political-organisational scales (e.g. sub-national, national, international), societal/governmental functions (e.g. production and consumption; policy-making, implementation and delivery), societal domains (e.g. public sector, market, civil society), infrastructure sectors (e.g. housing, energy, transport, water, waste) or policy-making arenas (Hodson and Marvin, 2012). What they have in common is that they perform 'bridging'/connecting and brokering/interest-intermediation roles among poorly connected actors or groups (Bush et al., 2017). As such what defines intermediaries is their position relative to other actors: they are situated in 'structural holes', for example between well-connected actor groups. Intermediaries are therefore part of the (governance) network rather than 'governing' or 'managing' it from a distance.

Instead of taking 'complexity' as a starting point, governance networks here are seen as vehicles to counteract structural-organisational fragmentation and siloedness that resulted from the privatisation of assets and the deregulation and liberalisation of markets – for example in the case of infrastructures – as part of the managerial reforms aimed at refocusing governments' remit to a few core functions (Guy et al., 2011) According to the intermediation approach, some form of integration among disconnected actors and groups is necessary to respond to contemporary environmental, economic and social pressures due to interdependency. Consequently, there is an emphasis on (network) structure and effective and efficient flows and pooling of information and resources. The role of intermediaries in turn is to 'facilitate relationships between key actors and enable sharing and pooling of knowledge' (Bush et al., 2017, p. 138) and resources to create capacity and momentum for transformative change.

Hodson and Marvin (2010) contend that intermediary organisations require specific resources themselves in order to be able to operate successfully. These include financial and human resources, established powers, appropriate (internal) organisational structures and cultures, relevant knowledge base and competence among staff, platforms for communication and opportunities for developing local contextual presence and visibility. Such resources are necessary because, as the relational-structural perspective highlights, intermediaries do not operate in a vacuum. As integral parts of the network of organisations, their possibilities for intermediation are influenced by interactions patterns in the network. As such, the fragmented 'multilevel' governance context which creates the need for intermediation also constrains the options of intermediaries to successfully perform their tasks. In addition, the way in which the intermediary is embedded in a governance context also influences its potential for facilitating change (Hawkey, Webb and Winskel, 2013; Bush et al., 2017).

The two approaches described above (TM, intermediation), both provide descriptive analytical frameworks of *what* low-carbon transitions in urban systems entail, as well as prescriptive frameworks of *how* they ought to be facilitated through active steering and involvement in the network. In terms of the 'what' question, the arguments are rooted in a 'default' characterisation of urban transitions as complex problems in a surrounding fragmented organisational environment. In terms of conceptualising networks and network processes (the 'how' question), the main focus is on 'collaboration' and 'coordination' – similarly to the inter-organisational and inter-governmental perspectives on networks of governance. However, references to politics and asymmetric power relations among network actors in different cities, and the potential for network links to convey 'influence' (as in the policy network perspective) are limited. This may be problematic because

> [b]eyond the large-scale social and economic forces usually identified by the literature as the drivers of network governance other local and/or agency factors come into play, such as political leadership, political and corporate cultures, institutional frameworks and social mobilisations.
>
> (Blanco, 2015, p. 125)

In addition, both TM and the intermediation model involve a form of network steering or coordination (transition manager, intermediary organisation) – which has implications in terms of network structure. Studies describing applications of the frameworks, however, seldom include a consideration for the implications of the structural features of (existing and/or new) networks, including the position of the transition manager and intermediary organisations within. Instead, the focus is on 'assigned' powers for management and intermediation.

Both frameworks were originally developed for facilitating low-carbon transitions in specific sectors, and were subsequently 'rescaled' and applied to the urban scale. In contrast with their focus on strategic interventions aimed at restructuring (urban) governance networks, empirical investigations of attempts at governing urban climate change mitigation found that these tended to develop on a largely ad hoc basis when and where opportunities emerged – rather than being driven by an integrated, strategic approach (Bulkeley and Broto, 2013; Nagorny-Koring and Nochta, 2018). These studies uncovered a patchwork of parallel, piecemeal responses – some of which do not rely on network governance:

- self-governing (emissions reduction from the municipalities' own estates and operations);
- provision (shaping local practices through providing or supplying particular services or resources);
- regulation (based on traditional local state authority);
- enabling (facilitating and encouraging action in the private sphere); and
- partnerships (co-producing initiatives with other stakeholders).

These observations highlight the importance of these piecemeal responses and consider them as governance 'experiments' to develop new modes of governing urban sustainability transitions in a more bottom-up, organic manner. After all, the complexity and uncertainty of sustainable urban transitions may be perceived in different ways in diverse cultural and institutional-organisational settings, characterised by varying levels of fragmentation (of knowledge, powers, authority and responsibilities) and different local rationales for transitions. In addition, 'transitions in urban systems are as likely to emerge from the co-incident actions of multiple agents and everyday actions as from purposive attempts to transform the city' (Bulkeley et al., 2011, p. 5).

Consequently, there is a need to deepen our understanding of the niche roles that network forms of governance can (or should) play in urban low-carbon transitions in different cities – alongside other responses as identified above. This involves paying more attention to flows of influence (politics and power), and issues of exclusion and inclusion, among actors in addition to processes of collaboration and coordination. There is furthermore a need to better understand the role and potential of Local Authorities (who have a mandate to govern cities and urban development with, and on behalf of, citizens) in facilitating and coordinating low-carbon transitions as part of existing and/or new, emerging networks. This requires an operationalisation of the network concept not only as a metaphor but also as an analytical tool to unravel the existing relationships between actors. Ultimately, new networks are interventions into the existing patterns of relationships which have implications for the outcomes produced – in this case, urban low-carbon development.

Setting out the research agenda

Three themes emerge requiring further elaboration and investigation in order to better understand the potential and role of network forms of governance in urban low-carbon transitions. First, the impact of 'context' on governance networks, including local historical and cultural factors. Second, the conceptualisation of 'networks' in urban low-carbon transitions: as a metaphor (as it often appears in the existing literature), or also as a useful analytical tool. Third, the implications of differences and similarities among low-carbon transition networks in different cities on network impact – in other words, the different roles that networks may play in transitions in different contextual settings. These are the three key foci around which the discussions presented throughout this book are centred.

The following chapters outline, apply and test a methodology to assess the implications of contextual differences for networks in the governance of urban low-carbon transitions. This methodology builds on both quantitative organisational network analysis (Chapter 2), as well as qualitative analyses of urban low-carbon energy transitions and their governance (Chapters 3, 4 and 5). The central aim of this mixed methods research is to examine whether the network concept as an analytical tool can provide useful insights via visualising local low-carbon transition networks and analysing their structure. Such insights regard the similarities and differences among the existing networks in the different cities, as well as an initial high-level overview to help interpret in-depth qualitative information. Qualitative data in turn is used first, to uncover and describe how local contextual characteristics may influence the networks – resulting in distinct path dependencies and accounting for the differences and similarities among the network structures. Second, to assess the networks' role and potential to facilitate low-carbon transitions. As such, the study deploys a concurrence research strategy by using the different methods to gather different types of insights and answer different sets of questions (Marcussen and Olsen, 2007).

The emergence of the network concept in policy-making and implementation was discussed in a historical context of waves of reorganisation processes within the public sector and in terms of its position and role relative to the private domain. This overview highlighted how the traditional 'Western' welfare state paradigm, associated with public sector bureaucracies and hierarchical command-and-control mechanisms, has been challenged by new ways of thinking about the 'ideal' role of governments in society starting from the late 1970s. The perceived successes of market competition in terms of efficiency gains facilitated a wave of reforms aimed at making the functioning of the public sector more 'business-like'. Arguably, the ways in which these changes played out in different countries is important not only to understand the growing interest in network-type concepts in public policy and administration research and practice but also to explain why these concepts may be interpreted and operationalised in different ways in different places.

One reason for this is that the reforms have not formed a coherent, monolithic approach or theory – rather, 'New Public Management' and 'managerial reforms' have been used as umbrella terms referring to a set of concepts and principles, and a toolbox of techniques. Ultimately, different countries interpreted and implemented these selectively, resulting in place-specific combinations with existing cultures and structures of governing (Pollitt and Bouckaert, 2011). In addition, a growing evidence base demonstrates that the 'transformation' of the public sector *in practice* has likely been more partial – even in the pioneering countries such as the United Kingdom or the United States – than what had previously been assumed (Lynn, 2011; Pollitt and Bouckaert, 2011).

These are important findings to consider for investigating the role of networks and networked forms of governance in facilitating and delivering low-carbon transitions. As network governance appears *alongside* the already functioning hierarchical and market-style arrangements, rather than replacing them altogether, differences in how these existing systems function are also likely to have implications for implementing network forms of governance (Christensen and Lægreid, 2011; Skelcher, Sullivan and Jeffares, 2013). First, this means that hierarchical relationships and vertical links are likely to exist between network actors as authority, powers and resources are not necessarily evenly distributed, and access to the network may be restricted. Power differences and issues of inclusion and exclusion of certain actors and interests are in turn important factors in the evolution of the framing of problems and solutions in governance networks (Klijn and Koppenjan, 2015). Second, diverse locationally specific (urban) governance arrangements may provide different opportunities and constraints for networks and networked forms of governance to become embedded in decision-making processes – and contribute to facilitating and advancing low-carbon transitions. Different network settings have also been shown to give way to different types of processes (information exchange, resource pooling, exerting influence); however, analyses of this potential diversity, and its implications, are lacking.

In order to unpack these issues and deepen our understanding of the impact of 'context' on the functioning and role of governance networks in urban low-carbon transitions, a variety of factors need to be examined. These factors that influence the emergence of network forms of governance have been summarised by Torfing (2005) as follows:

1 societal and organisational fragmentation (for example, as a result of reforms and the dispersion of powers and authority);
2 complexity in contemporary decision-making due to uncertainty (due to insufficient understanding and lack of societal consensus over the definitions, causes and effects of problems and solution options); and

3 new societal dynamics related to a shift from 'government' to 'governance' (relating to changes in perceptions of the 'appropriate' roles that the state – national and local –,businesses and communities or individual citizens can and should play in societal coordination).

The key argument is that the ways in which these three processes played out historically in different cities can provide important insights regarding local interpretations and appropriateness of the role of networks and networked forms of governance. As such, analyses focusing on these key factors can help us to understand whether or not these differences exist in reality, the impact they have on networks in governance, and on the potential contribution of networks and horizontal coordination to low-carbon transitions in different cities. Therefore, the case studies presented in Chapters 3, 4 and 5 follow this framework to investigate the emergence of networks, and the associated implications for local low-carbon transition trajectories in the energy sector. The case studies focus on the energy transition networks in three European cities: Birmingham (UK), Frankfurt (Germany) and Budapest (Hungary).

References

Anguelovski, I. and Carmin, J. A. (2011) 'Something borrowed, everything new: Innovation and institutionalization in urban climate governance', *Current Opinion in Environmental Sustainability*, 3(3), pp. 169–175. doi:10.1016/j.cosust.2010.12.017.

Avelino, F. (2009) 'Empowerment and the challenge of applying transition management to ongoing projects', *Policy Sciences*, 42(4), pp. 369–390. doi:10.1007/s11077-009-9102-6.

Bai, X. (2007) 'Integrating global environmental concerns into urban management: The scale and readiness arguments', *Journal of Industrial Ecology*, 11(2), pp. 15–29. doi:10.1162/jie.2007.1202.

Bevir, M. (ed.) (2011) 'Governance as theory, practice, and dilemma', in *The SAGE Handbook of Governance*, pp. 1–16. SAGE. doi:10.4135/9781446200964.n1.

Bevir, M. and Rhodes, R. A. W. (2016) *Rethinking Governance: Ruling, Rationalities and Resistance.* Routledge. doi:10.4324/9781315712949.

Bevir, M. and Richards, D. (2009) 'Decentring policy networks: A theoretical agenda', *Public Administration*, 87(1), pp. 3–14. doi:10.1111/j.1467-9299.2008.01736.x.

Blanco, I. (2015) 'Between democratic network governance and neoliberalism: A regime-theoretical analysis of collaboration in Barcelona', *Cities*, 44, pp. 123–130. doi:10.1016/j.cities.2014.10.007.

Börzel, T. A. (1998) 'Organizing Babylon: On the different conceptions of policy networks', *Public Administration*, 76(2), pp. 253–273. doi:10.1111/1467-9299.00100.

Broto, V. C. and Bulkeley, H. (2013) 'A survey of urban climate change experiments in 100 cities', *Global Environmental Change*, 23(1), pp. 92–102. doi:10.1016/j.gloenvcha.2012.07.005.

Bulkeley, H. (2000) 'Discourse coalitions and the Australian climate change policy network', *Environment and Planning C: Government and Policy*, 18(6), pp. 727–748. doi:10.1068/c9905j.

Bulkeley, H.*et al.* (2011) 'Introduction', in H. Bulkeley*et al.* (eds), *Cities and Low Carbon Transitions*, pp. 1–10. Routledge. Available at: www.europeanfinancialreview.com/?p=3541 (Accessed: 28 February 2020).

Bulkeley, H. and Betsill, M. M. (2005) 'Rethinking sustainable cities: Multilevel governance and the "urban" politics of climate change', *Environmental Politics*, 14(1), pp. 42–63. doi:10.1080/0964401042000310178.

Bulkeley, H. and Betsill, M. M. (2013) 'Revisiting the urban politics of climate change', *Environmental Politics*, 22(1), pp. 136–154. doi:10.1080/09644016.2013.755797.

Bulkeley, H. and Broto, V. C. (2013) 'Government by experiment? Global cities and the governing of climate change', *Transactions of the Institute of British Geographers*, 38(3), pp. 361–375. doi:10.1111/j.1475-5661.2012.00535.x.

Bulkeley, H., Broto, V. C. and Edwards, G. A. S. (2015) *An Urban Politics of Climate Change: Experimentation and the Governing of Socio-Technical Transitions.* Routledge. doi:10.1017/CBO9781107415324.004.

Bush, R. E.*et al.* (2017) 'The role of intermediaries in low carbon transitions: Empowering innovations to unlock district heating in the UK', *Journal of Cleaner Production*, 148, pp. 137–147. doi:10.1016/j.jclepro.2017.01.129.

Christensen, T. and Lægreid, P. (2011) 'Complexity and hybrid public administration: Theoretical and empirical challenges', *Public Organization Review*, 11(4), pp. 407–423. doi:10.1007/s11115-010-0141-4.

Drechsler, W. (2005) 'The rise and demise of the new public management', *Post-Autistic Economics Review*, 33 (14), pp. 17–28.

Dunleavy, P. and Hood, C. (1994) 'From old public administration to new public management', *Public Money and Management*, 14(3), pp. 9–16. doi:10.1080/09540969409387823.

European Committee of the Regions (2018) COP24: Cities and regions call for a formal role in the Paris Agreement. Available at: https://cor.europa.eu/en/news/Pages/COP24-cities-and-regions-call-for-a-formal-role-in-the-Paris-Agreement.aspx (Accessed: 26 February 2020).

Frantzeskaki, N., Wittmayer, J. and Loorbach, D. (2014) 'The role of partnerships in 'realising' urban sustainability in Rotterdam's city ports area, the Netherlands', *Journal of Cleaner Production*, 65, pp. 406–417. doi:10.1016/j.jclepro.2013.09.023.

Gerlak, A. K. (2015) 'Resistance and reform: Transboundary water governance in the Colorado river delta', *Review of Policy Research*, 32(1), pp. 100–123. doi:10.1111/ropr.12114.

Grin, J., Rotmans, J. and Loorbach, D. (2010) 'Research into the governance of transitions: A framework for transition management', in J. Schot, J. Grin and J. Rotmans (eds), *Transitions to Sustainable Development: New Directions in the Study of Long Term Transformative Change*, pp. 140–160. Routledge. doi:10.4324/9780203856598.

Guy, S., Marvin, S., Medd, W. and Moss, T. (eds) (2011) *Shaping Urban Infrastructures: Intermediaries and the Governance of Socio-Technical Networks.* Routledge.

Hajer, M. (2003) 'Policy without polity? Policy analysis and the institutional void', *Policy Sciences*, 36(2), pp. 175–195. doi:10.1023/A:1024834510939.

Hawkey, D., Webb, J. and Winskel, M. (2013) 'Organisation and governance of urban energy systems: District heating and cooling in the UK', *Journal of Cleaner Production*, 50, pp. 22–31. doi:10.1016/j.jclepro.2012.11.018.

Hodson, M. and Marvin, S. (2010) 'Can cities shape socio-technical transitions and how would we know if they were?', *Research Policy*, 39 (4), pp. 477–485. doi:10.1016/j.respol.2010.01.020.

Hodson, M. and Marvin, S. (2012) 'Mediating low-carbon urban transitions? Forms of organization, knowledge and action', *European Planning Studies*, 20(3), pp. 421–439. doi:10.1080/09654313.2012.651804.

Hodson, M., Marvin, S. and Bulkeley, H. (2013) 'The intermediary organisation of low carbon cities: A comparative analysis of transitions in Greater London and Greater Manchester', *Urban Studies*, 50(7), pp. 1403–1422. doi:10.1177/0042098013480967.

Hölscher, K.*et al.* (2019) 'Opening up the transition arena: An analysis of (dis) empowerment of civil society actors in transition management in cities', *Technological Forecasting and Social Change*, 145, pp. 176–185. doi:10.1016/j.techfore.2017.05.004.

Homburg, V., Pollitt, C. and van Thiel, S. (2007) '*Introduction*', in C. Pollitt, S. van Thiel and V. Homburg (eds), *New Public Management in Europe*, pp. 1–9. Palgrave Macmillan. doi:10.1057/9780230625365_1.

Kapucu, N.*et al.* (2010) 'Interorganizational network coordination under stress caused by repeated threats of disasters', *Journal of Homeland Security and Emergency Management*, 7(1). doi:10.2202/1547-7355.1629.

Kapucu, N. and Demiroz, F. (2017) 'Interorganizational networks in disaster management', in E. C. Jones and A. J. Faas (eds), *Social Network Analysis of Disaster Response, Recovery, and Adaptation*, pp. 25–39. Butterworth-Heinemann. doi:10.1016/B978-0-12-805196-2.00003-0.

Kern, F. and Smith, A. (2008) 'Restructuring energy systems for sustainability? Energy transition policy in the Netherlands', *Energy Policy*, 36(11), pp. 4093–4103. doi:10.1016/j.enpol.2008.06.018.

Kersbergen, K. V. and Waarden, F. V. (2004) '"Governance' as a bridge between disciplines: Cross-disciplinary inspiration regarding shifts in governance and problems of governability, accountability and legitimacy', *European Journal of Political Research*, 43(2), pp. 143–171. doi:10.1111/j.1475-6765.2004.00149.x.

Kivimaa, P.*et al.* (2019) 'Towards a typology of intermediaries in sustainability transitions: A systematic review and a research agenda', *Research Policy*, 48 (4), pp. 1062–1075. doi:10.1016/j.respol.2018.10.006.

Klijn, E. H. and Koppenjan, J. (2012) 'Governance network theory: Past, present and future', *Policy and politics*, 40(4), pp. 587–606. doi:10.1332/030557312X655431.

Klijn, E. H. and Koppenjan, J. (2015) *Governance Networks in the Public Sector*. Routledge. doi:10.4324/9781315887098.

Klijn, E. H. and Skelcher, C. (2007) 'Democracy and governance networks: Compatible or not', *Public Administration*, 85(3), pp. 587–608. doi:10.1111/j.1467-9299.2007.00662.x.

Lewis, J. M. (2011) 'The future of network governance research: Strength in diversity and synthesis', *Public Administration*, 89(4), pp. 1221–1234. doi:10.1111/j.1467-9299.2010.01876.x.

Lewis, J. M., Baeza, J. I. and Alexander, D. (2008) 'Partnerships in primary care in Australia: Network structure, dynamics and sustainability', *Social Science and Medicine*, 67(2), pp. 280–291. doi:10.1016/j.socscimed.2008.03.046.

LGMA (no date) About the LGMA – Cities & Regions Pavilion. Available at: www.cities-and-regions.org/about-the-lgma/ (Accessed: 26 February 2020).

Loorbach, D. (2010) 'Transition management for sustainable development: A prescriptive, complexity-based governance framework', *Governance*, 23(1), pp. 161–183. doi:10.1111/j.1468-0491.2009.01471.x.

Lynn, L. E. (2011) 'The persistence of hierarchy', in M. Bevir (ed.), *The SAGE Handbook of Governance*, pp. 218–236. SAGE.

Marcussen, M. and Olsen, H. P. (2007) 'Transcending analytical cliquishness with second-generation governance network analysis', in J. Torfing and M. Marcussen (eds), *Democratic Network Governance in Europe*, pp. 273–292. Palgrave Macmillan. doi:10.1057/9780230596283_14.

Markard, J., Raven, R. and Truffer, B. (2012) 'Sustainability transitions: An emerging field of research and its prospects', *Research Policy*, 41(6), pp. 955–967. doi:10.1016/j.respol.2012.02.013.

Mayntz, R. (2019) 'From government to governance', in D. Scheer and F. Rubik (eds), *Governance of Integrated Product Policy*, pp. 18–25. Routledge. doi:10.4324/9781351282604-2.

McCarthy, M. P., Best, M. J. and Betts, R. A. (2010) 'Climate change in cities due to global warming and urban effects', *Geophysical Research Letters*, 37 (9). doi:10.1029/2010GL042845.

Mcgranahan, G. and Satterthwaite, D. (2014) Urbanisation concepts and trends. Available at: https://pubs.iied.org/pdfs/10709IIED.pdf (Accessed: 28 February 2020).

Meuleman, L. (2008) *Public Management and the Metagovernance of Hierarchies, Networks and Markets.* Physica-Verlag HD. doi:10.1007/978-3-7908-2054-6.

Nagorny-Koring, N. C. and Nochta, T. (2018) 'Managing urban transitions in theory and practice: The case of the Pioneer Cities and Transition Cities projects', *Journal of Cleaner Production*, 175, pp. 60–69. doi:10.1016/j.jclepro.2017.11.072.

Nevens, F.*et al.* (2013) 'Urban transition labs: Co-creating transformative action for sustainable cities', *Journal of Cleaner Production*, 50, pp. 111–122. doi:10.1016/j.jclepro.2012.12.001.

Nochta, T. and Skelcher, C. (2020) 'Network governance in low-carbon energy transitions in European cities: A comparative analysis', *Energy Policy*, 138. doi:10.1016/j.enpol.2020.111298.

Osborne, S. P. (2010) *The New Public Governance?: Emerging Perspectives on the Theory and Practice of Public Governance.* Routledge. doi:10.4324/9780203861684.

Pollitt, C. (2003) 'Public management reform', *OECD Journal on Budgeting*, 3 (3), pp. 121–134. doi:10.1787/budget-v3-art19-en.

Pollitt, C. and Bouckaert, G. (2011) *Public Management Reform – A Comparative Analysis: New Public Management, Governance, and the Neo-Weberian State*, 3rd edn. Oxford University Press.

Provan, K. G. and Kenis, P. (2008) 'Modes of network governance: Structure, management, and effectiveness', *Journal of Public Administration Research and Theory*, 18(2), pp. 229–252. doi:10.1093/jopart/mum015.

Provan, K. G. and Milward, H. B. (2001) 'Do networks really work? A framework for evaluating public-sector organizational networks', *Public Administration Review*, 61(4), pp. 414–423. doi:10.1111/0033-3352.00045.

Rhodes, R. A. W. (1996) 'The new governance: Governing without government', *Political Studies*, 44(4), pp. 652–667. doi:10.1111/j.1467-9248.1996.tb01747.x.

Rhodes, R. A. W. (1997) *Understanding Governance: Policy Networks, Governance, Reflexivity and Accountability.* Open University Press.

Scharpf, F. W. (2001) 'Notes toward a theory of multilevel governing in Europe', *Scandinavian Political Studies*, 24(1), pp. 1–26. doi:10.1111/1467-9477.00044.

Skelcher, C. (2000) 'Changing images of the state: Overloaded, hollowed-out, congested', *Public Policy and Administration*, 15(3), pp. 3–19. doi:10.1177/095207670001500302.

Skelcher, C., Sullivan, H. and Jeffares, S. (2013) *Hybrid Governance in European Cities: Neighbourhood, Migration and Democracy.* Springer.

Smith, A., Stirling, A. and Berkhout, F. (2005) 'The governance of sustainable socio-technical transitions', *Research Policy*, 34(10), pp. 1491–1510. doi:10.1016/j.respol.2005.07.005.

Torfing, J. (2005) 'Governance network theory: Towards a second generation', *European Political Science*, 4(3), pp. 305–315. doi:10.1057/palgrave.eps.2210031.

Torfing, J. (2012) 'Governance Networks', in *The Oxford Handbook of Governance*. Oxford University Press. doi:10.1093/oxfordhb/9780199560530.013.0007.

UNFCCC (2015) The Paris Agreement. Available at: https://unfccc.int/process-and-meetings/the-paris-agreement/the-paris-agreement (Accessed: 26 February 2020).

Voytenko, Y.*et al.* (2016) 'Urban living labs for sustainability and low carbon cities in Europe: Towards a research agenda', *Journal of Cleaner Production*, 123, pp. 45–54. doi:10.1016/j.jclepro.2015.08.053.

Wittmayer, J. M.*et al.* (2018) 'Transition Management: Guiding Principles and Applications', in N. Frantzeskaki*et al.* (eds) *Co-creating Sustainable Urban Futures*, pp. 81–101. Springer. doi:10.1007/978-3-319-69273-9_4.

2 Analysing energy transition networks in Birmingham, Frankfurt and Budapest

Introduction

This chapter provides a comparative analysis of the existing energy transition networks in Birmingham, Frankfurt and Budapest, to enable potential similarities and differences among the three cases to be identified. The analysis also provides a high-level overview of the existing participative decision-making arenas in each city, the actors involved and their interaction patterns. This information is used to reconstruct the existing energy transition networks in the three cities. Network-structural analysis and statistical measures are used to describe network topographies, the types of actors involved, and their potential to influence decision-making processes based on their position within the network relative to other actors. These initial insights will be subject to more in-depth investigation, and validation, through qualitative analyses in Chapters 3, 4 and 5.

(Re)Constructing the energy transition networks

The selection of city cases

The policy focus of this investigation is the energy sector (excluding transport). The decarbonisation of energy systems, involving production, supply and demand, is a central challenge for sustainable development and climate change mitigation (Rockström et al., 2017). Inspired by this low-carbon agenda, many cities aspire to gain more oversight and control over energy production, supply and demand, resulting in a growing interest in decentralised energy schemes (Transition Cities, 2015). Examples include combined heat and power (CHP) plants, waste-to-energy schemes or solar power plants (IEA, 2017). Although energy systems governance is traditionally nationally organised in most countries, these decentralised schemes provide opportunities for Local Authorities and other local actors to enter the national regimes of energy supply.

The selection of city cases for this study has been driven by a 'most different cases' approach (Bryman, 2016), within a common European context. The underlying assumption was that while the largely different contexts may give opportunity to reveal disjunctures and differences regarding the role and potential of governance networks to support low-carbon energy transitions, some commonality in terms of the European political-institutional framework is expected to enable meaningful comparison. Further, the case selection was also informed by previous research conducted within the frames of the 'Transition Cities' project (Climate-KIC, no date).

Three cities are chosen for comparative analysis of existing low-carbon energy transition networks: Birmingham, United Kingdom; Budapest, Hungary; and Frankfurt, Germany. Based on previous research (Kuhlmann and Wollmann, 2014; Swianiewicz, 2014; Copus, Roberts and Wall, 2017), these cities represent diverse local settings in terms of:

- state structures and national contexts (unitary-centralised, unitary-decentralised and federal-decentralised respectively);
- degrees of local government legal and fiscal authority (weak/moderate, moderate and strong comparatively); and
- Local Authority roles in energy systems governance (little/no involvement, heat production/supply only, heat and electricity production/supply).

Whilst a variety of low-carbon projects are underway in the cities chosen for in-depth analysis, the Transition Cities project found that transitions in energy systems (electricity and heat) are key consideration for the participating cities, including Birmingham, Budapest and Frankfurt (Transition Cities, 2015).

The methodology of reconstructing and analysing energy transition networks

The aim of the network analysis in this study is to develop a relational perspective on social phenomena, representing it as a structure consisting of 'nodes' (organisations or individuals) linked to one another through network 'ties' (Prell, 2012). In this study, network analysis is used to map the existing participative, multi-actor governance initiatives (commissions, advisory groups, working groups, roundtables) in each city, and the organisations involved in these (i.e. municipal departments, businesses, co-operatives, civil sector organisations, etc.). Network visualisations are therefore based on membership data, represented as organisational 'sociograms' (network graphs).

The underlying assumption is that such multi-actor decision-making initiatives create space for inter-organisational interactions, and that the emerging interaction patterns within and between these initiatives contribute to the development of governance networks over time. Such developing network structures in turn are likely to enable and/or constrain flows of influence, information and resources and therefore, according to the network logic of governance, possess explanatory power with regards to outcomes – in this research, the pace and trajectory of low-carbon energy transitions (Marcussen and Olsen, 2007).

Data collection for this study took place over the course of approximately 18 months between 2015 and 2017. The multi-actor arenas which were used to reconstruct the energy transition networks in each city were those with a remit that in some form included local low-carbon energy. Information collection regarding existing participative and collaborative decision-making arenas, and their membership, was facilitated through previous research (Transition Cities project), web searches and municipal publications and publicly available agendas and meeting minutes. In a small number of instances, in-depth interviews with city officials and local stakeholders were also used to complement transition arena memberships, specifically in instances where the available information gathered from secondary sources was unclear or outdated (for a note on the qualitative data collection and a list of interviewees see Appendix I). Table 2.1 presents the surveyed decision-making arenas in each city.

The data on organisations' membership in the different arenas was used to produce network visualisations as two-mode (bipartite) graphs – depicting organisations' involvement in different multi-actor decision-making initiatives (sec also Nochta and Skelcher, 2020). The network is called 'two-mode' because two types of nodes are present in the network graphs: decision-making arenas and member organisations. Each organisation (node type 1) is connected to the arena(s) (node type 2) in which they participate through a network tie (edge) (Borgatti and Halgin, 2011). Such two-mode networks allow for basic actor analysis and provide information on the size and density of the network, as well as overlaps between different decision-making arenas in terms of their membership.

Table 2.1 Lists of the surveyed multi-actor decision-making arenas in the three cities

Birmingham	*Frankfurt*	*Budapest*
Green Commission (GC)	Environmental Forum/ Umweltforum Rhein-Main	Urban Development Plans (Integrated Urban Development Plan and Strategy, Budapest 2030 Plan)
GC Green Growth Roundtable	Frankfurt Green City Sustainability Board	Environmental Programs
GC Energy & Resources Roundtable	M100CP Advisory Board/ Klimaschutzbeirat	Inventories of the State of the Environment / SEAP
GC Buildings & Efficiency Roundtable	Industry Masterplan Advisory Board	nnovative Services Company (ISCO) Advisory Group
GC Transport & Mobility Roundtable	Industry Masterplan Energy Working Group	Compact of Mayors/ Under 2 Steering Group
GC Natural Capital & Adaptation Roundtable	Regional Energy Concept – Board	
Sustainability West Midlands (SWM)	Regional Energy Concept – Energy Working Group	
SWM Board	Regional Energy Concept – Economy Working Group	
Birmingham Science City Innovative Low Carbon Working Group (BSC LCWG)	Regional Energy Concept – Mobility Working Group	
Greater Birmingham And Solihull Local Enterprise Partnership (GBS LEP)	Regional Energy Concept – Housing Working Group	
GBS LEP Business Board	Regional Energy Concept – Finance Working Group	

However, developing a better understanding of the relations between different organisations requires the transformation of two-mode graphs into one-mode ones by eliminating the collaborative decision-making arenas. One-mode networks helps to analyse organisations' relations to one another in a more direct way. Here, network nodes, representing the organisations, are linked to one another based on mutual participation in the same decision-making arena(s). Simple network statistical measures were used to obtain information regarding the network as a whole (size, density and centralisation) as well as the position of certain nodes relative to the network based on different measures of node centrality (degree, betweenness and closeness). Details of the network statistical measures used in this study are described in Table 2.2.

Table 2.2 Network statistical measures for analysing the structural characteristics of energy transition networks

Network measure	*Definition*	*Information provided*
Network size	Number of nodes in the network.	Number of decision-making arenas and of the actors involved, used as proxy for outreach to stakeholders.
Density	Proportion of ties present in the network compared to the possible total number of ties.	Overlap between decision-making arenas in terms of membership, related to the interconnectedness of the network.
Centralisation	The extent (proportion) to which one actor in the network holds all ties present in the network.	Indicates actor groups centralised around another actor; to determine whether high density results from general interconnectedness, or from the presence of a few well-connected actors.
Degree centrality	Number of immediate contacts owned by any one actor, i.e. the number of ties/edges belonging to a node.	To identify the most well-connected actors who are involved in multiple arenas.
Betweenness centrality	Calculated on the basis of how many times an actor (node) sits on the shortest path between two unconnected actors.	To identify actors who connect separate groups and play a bridging/brokerage role in the network.
Closeness centrality	Measuring 'independence' of a node, calculated from the distance between actors; actors positioned to the shortest distance to others have higher scores.	To identify actors who have the greatest potential to influence the network processes.

Source: Prell, 2012.

The reconstruction of network graphs and the statistical analyses are conducted using Gephi[1] software. Gephi has a variety of in-built methods for generating intuitive graph layouts. To create the network graphs presented here, two of these are used: first, the 'Fruchterman Reingold' and in a second step, the 'Force Atlas 2' layout algorithm (Grandjean, 2015). The 'MultiMode Networks Transformations'[2] plugin allows for transforming two-mode graphs into one-mode ones. All network files (one- and two-mode) are accessible online[3] (Gephi format). Examples of the produced one-mode network visualisations are shown in Figures 2.1–2.3. Nodes in the network are assigned different colours on the grey scale based on their sectoral origin (public and market sector, community/third sector and academia/research).

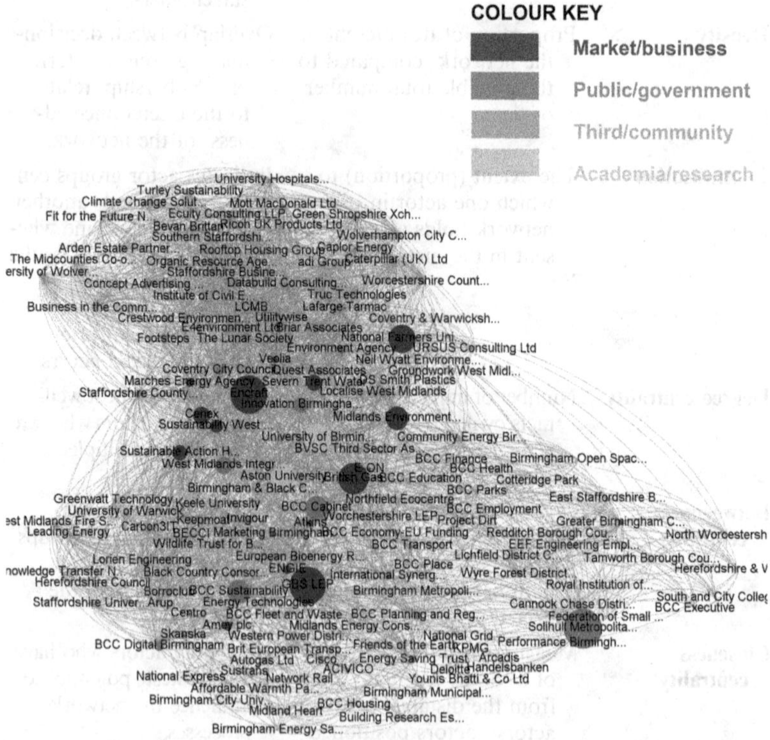

Figure 2.1 The energy transition network in Birmingham – One-mode network visualisation, whole network (node sizes determined by betweenness centrality scores).

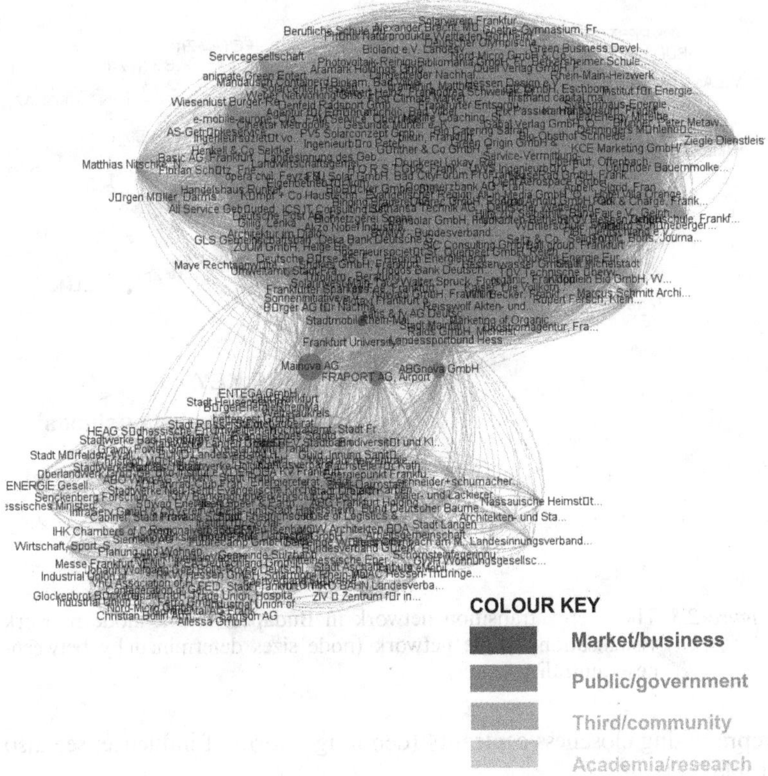

Figure 2.2 The energy transition network in Frankfurt – One-mode network visualisation, whole network (node sizes determined by betweenness centrality scores).

Node sizes in Figures 2.1–2.3 are determined by their betweenness centrality score (brokerage/intermediation role; see also Table 2.2).[4]

The network visualisations and statistical analyses generate comparable data across the three city cases through a systematic use of methods and techniques for data collection and analysis. As such they allow for comparisons to be made in terms of network size, structure and the types of actors involved. The comparative analysis extends to both the whole networks, as well as their potentially more influential core group (according to the relational understanding of 'power'). Core groups are made up of actors who participate in two or more decision-making arenas, and the network elations among them. Figures 2.4–2.6 show examples of core group visualisations from each city, with the node sizes

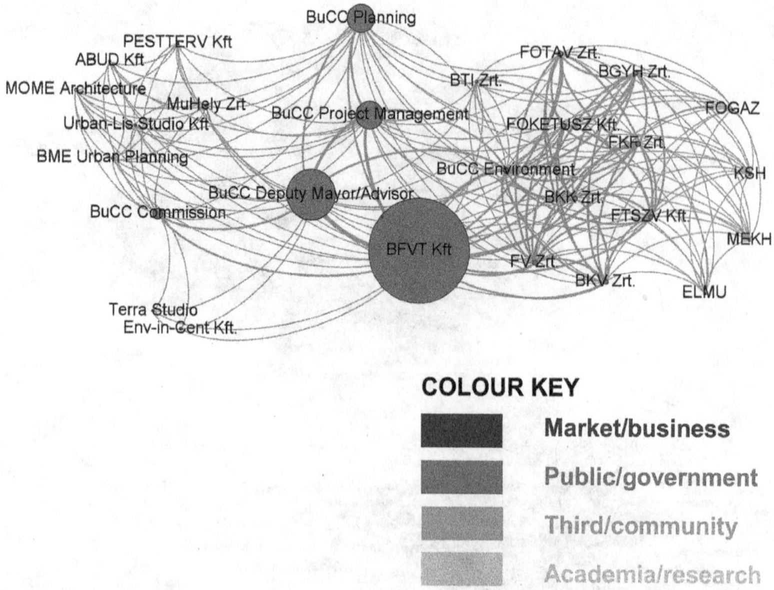

Figure 2.3 The energy transition network in Budapest – One-mode network visualisation, whole network (node sizes determined by betweenness centrality scores).

representing closeness centrality (denoting a form of influence; see also Table 2.2).

Comparative analysis of network structures

Network size

The data collection revealed the existence of a variety of formalised collaborative decision-making initiatives in two of the three cities – Birmingham and Frankfurt. In Budapest, network-type interactions were limited to semi-formal working groups which have been responsible for the production of various policy and progress monitoring documents published on a regular basis, commissioned by the City Council. Table 2.3 shows the network size, and the number of decision-making arenas and member organisations in the whole networks and their core groups in each city.

The Birmingham network consists of 11 decision-making arenas. Altogether these involve 148 member organisations. The core group, focusing on organisations that are involved in more than one

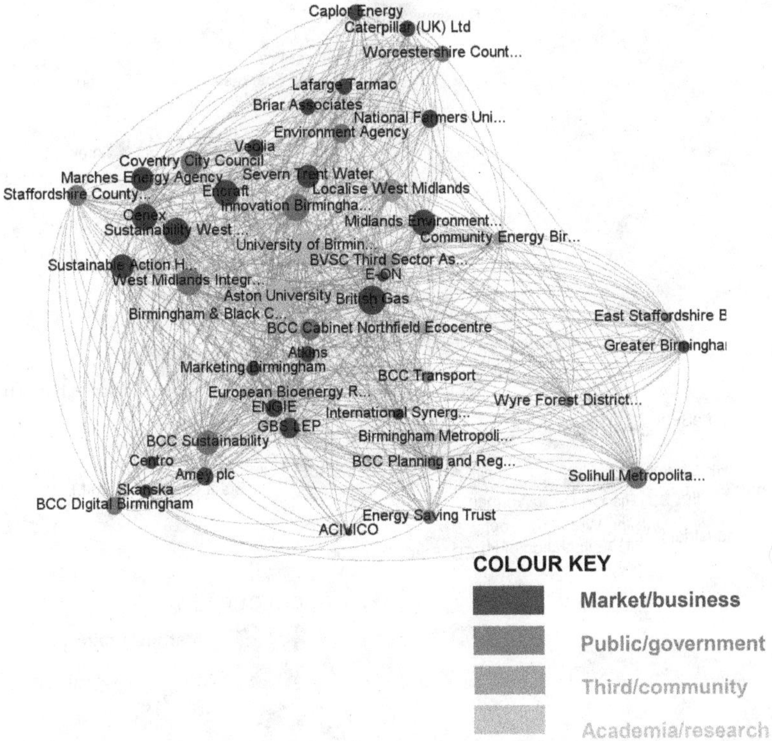

Figure 2.4 The energy transition network in Birmingham – One-mode network visualisation, core group (node sizes determined by closeness centrality scores).

decision-making arena, retains 33% of the entire network and consist of 49 member organisations. The largest network was found in Frankfurt which involves 270 organisations – considerably more than either Birmingham's or Budapest's network. Member organisations belong to 11 decision-making arenas. Many of the organisations, however, are members of just one arena, and therefore the core group involves only 17% of the entire network and consists of 47 organisations. The network found in Budapest is significantly smaller than Birmingham's or Frankfurt's. Furthermore, it is characterised by an absence of formalised collaborative decision-making arenas – the 5 arenas represented in the network visualisations are semi-formal working groups and collaborative initiative 'under

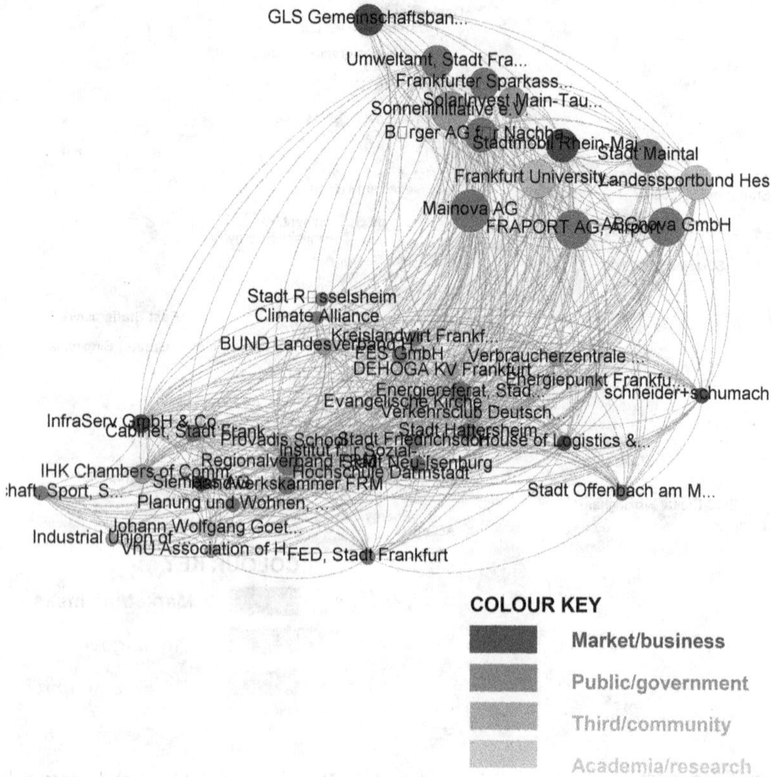

Figure 2.5 The energy transition network in Frankfurt – One-mode network visualisation, core group (node sizes determined by closeness centrality scores).

implementation' ('ISCO' – Innovative Services Company) – which is based on existing but ad-hoc interaction patterns among municipally owned utilities and companies. The core group involves more than half of the entire network (52%).

In terms of network governance processes, the analysis of the size of networks and their core groups reveals significant differences among the existing networks in Birmingham, Budapest and Frankfurt. Budapest's network appears comparatively weak and exclusive due to its semi-formal nature and comparatively small size. In Frankfurt, only a small number of network actors are involved in the core group of the network. This may represent significant separation, and potential for network-internal hierarchy to arise, between the more embedded (and therefore possibly more influential) core group and the network

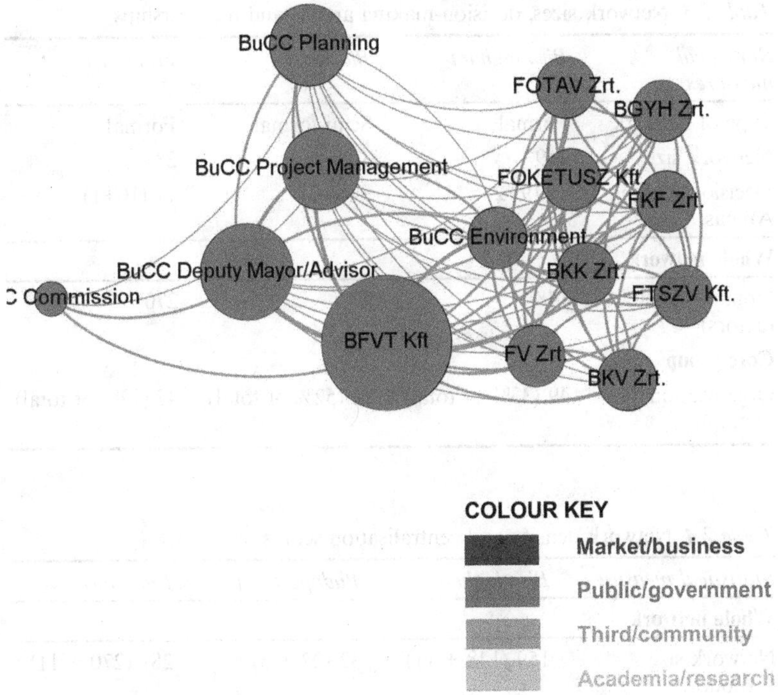

Figure 2.6 The energy transition network in Budapest – One-mode network visualisation, core group (node sizes determined by closeness centrality scores).

periphery. In contrast, the network structure in Birmingham appears more polycentric than in the other two cases, allowing for more horizontal relationships among network actors.

Network density and centralisation

Network density and centralisation measures are used to further investigate the structural characteristics of the networks from the three cities. They provide information on network cohesion and balance in terms of memberships of different decision-making arenas, respectively. Table 2.4 shows the density and centralisation scores for the whole networks and their core groups in our city cases (based on the two-mode networks).

In the case of the whole networks, Frankfurt's has the lowest density and the highest centralisation. This indicates less overlap between decision-making arena memberships, and that one (or more) arena has

Table 2.3 Network sizes, decision-making arenas and memberships

Statistical measures	*Birmingham*	*Budapest*	*Frankfurt*
Type of network	Formal	Semi-formal	Formal
Network size	159	32	281
Decision-making Arenas	11 (9+2)	5 (4+1)	11 (10+1)
Whole network			
Organisations (actors)	148	27	270
Core group			
Organisations (actors)	49 (33% of total)	14 (52% of total)	47 (17% of total)

Table 2.4 Network density and centralisation scores

Statistical measures	*Birmingham*	*Budapest*	*Frankfurt*
Whole network			
Network size (nodes)	159 (148 + 11)	32 (27 + 5)	281 (270 + 11)
Number of connections (ties)	263	47	363
Network density (%)	2.1	9.5	0.9
Centralisation score	0.32	0.38	0.55
Core group			
Network size (nodes)	60 (49 + 11)	19 (14 + 5)	59 (47 + 11)
Number of connections (ties)	164 (62% of total)	35 (73% of total)	140 (38% of total)
Network density (%)	9.5	20.5	8.5
Centralisation score	0.32	0.58	0.31

significantly more members than others in the network. This arena is the Environmental Forum / Umweltforum which is somewhat disconnected from the rest of the network but involves a large number of private sector stakeholders. Budapest's network appears to be tightly knit (high density) and relatively balanced. However, this can also be explained by the comparatively small network rather than a high degree of cohesion as network density is dependent on network size. Birmingham's network appears relatively well-connected and balanced according to the density and centralisation measures.

The core groups of the networks in Birmingham and Frankfurt seem fairly similar in terms of size, density and centralisation. Centrality remains the highest in Birmingham's core group. The sharp increase in density in Frankfurt's core group compared to the whole network confirms an apparent divide between the core and periphery of the network. Changes between the core group and the whole network in Budapest are less significant, strengthening the impression that the network is likely to be exclusive and closed.

Sectoral outreach and integration

The literatures on urban sustainability transitions emphasise the benefits of bringing together actors from various sectors (public, market and third sector, as well as academic/knowledge institutions) in deliberative, collaborative transition arenas. The underlying assumption here is that actors from different sectors possess diverse interests and resources which need to be brought together to produce visions of, and pathways towards, low-carbon city futures which have the support of diverse groups of actors.

The analysis of the sectoral makeup of the networks in Birmingham, Budapest and Frankfurt highlights further differences among the three cities (see Table 2.5). In particular, Budapest's network stands out as significantly different from the other two on the level of whole networks. The network is characterised by a high share of organisations with an origin in the public sector (Local Authority departments, publicly owned utilities and companies) which becomes even more prevalent in the core group. This indicates a relatively low level of integration of the private sector and communities into decision-making processes which is retained by the public sector in Budapest.

The networks in Birmingham and Frankfurt exhibit similarity in terms of sectoral share on the level of whole networks which also include actors who are only members in one decision-making arena. There is, however, a significant shift of sectoral shares in the core group of Frankfurt in contrary to Birmingham where the market sector appears to be dominant both in the whole network as well as the core group. In Frankfurt, the core group is dominated by public sector organisations instead of the market sector. This indicates a more central role for the public sector in the formulation and implementation of the local low-carbon energy agenda in Frankfurt. In contrast, sectoral

Table 2.5 Sectoral representation in the energy transition network and their core groups

	Birmingham	*Budapest*	*Frankfurt*
Whole network			
Public % (inc. publicly owned companies)	24%	67%	21%
Market % (private businesses)	55%	26%	60%
Third % (voluntary organisations & citizens' groups)	12%	0%	13%
Academia % (inc. private research institutes)	9%	7%	6%
Core group			
Public % (inc. publicly owned companies)	31%	100%	51%
Market % (private businesses)	52%	0%	13%
Third % (voluntary organisations & citizens' groups)	10%	0%	23%
Academia % (inc. private research institutes)	6%	0%	13%

shares remain broadly similar between the whole network and the core group in Birmingham, signalling higher levels of integration and the blurring of boundaries between the public and private domains. By extension, the private domain and specifically the market sector appears to have more influence of low-carbon transition trajectories in Birmingham than in the other two cities.

Actor analysis

Different types of network centrality measures are available for examining the position of individual actors within a network structure (Prell, 2012). To further analyse the energy transition networks of Birmingham, Budapest and Frankfurt, three network centrality measures are considered: degree, betweenness and closeness centrality (see also

Table 2.2). Degree centrality represents an actor's connectedness to others in the network ('popularity'), by simply counting the total number of connections (ties) they possess to others. Betweenness centrality aims to capture some form of 'bridging', brokerage or intermediary role of a node in a network. Nodes with high betweenness centrality are situated in 'structural holes' and therefore play a key role in keeping the network intact. Structural holes in network graphs denote areas of the network with a lack of links (ties) between two or more network components (nodes or well-connected groups of nodes; (Burt, 2017)). Betweenness centrality is calculated by counting the number of times a node is located on the shortest path between two other nodes. Closeness centrality is somewhat less intuitive than degree or betweenness centrality measures. It is aimed at capturing actors' 'independence', i.e. their ability to access and spread information and influence more quickly and directly to others in the network (Freeman, Roeder and Mulholland, 1979). Closeness centrality therefore builds on the importance of path lengths, and is measured based on distance between actors (relating to the number of nodes on the shortest path between one node and another).

All centrality scores are based on the one-mode network graphs which involve only one type of node: member organisations. In the one-mode graphs organisations are connected to one another through network ties if they participate in the same decision-making arena(s). Appendix II contains the centrality scores, according to the different centrality measures (degree, betweenness, closeness), of actors involved in the core groups of each of the three networks. Centrality scores for all other network actors can be obtained from the source files available online.[5] For the core groups, centrality scores are calculated both within the entire networks (including the links to actors at the network periphery) and also within the core groups (among more influential actors). Normalised scores are obtained by converting the number of links a node has into the fraction of its maximum possible level (Marsden, 2015). As such the highest possible normalised centrality score for any node (actor) is 1.

Degree and closeness centrality scores reveal a well-connected, integrated network in Birmingham with a high proportion of actors possessing more than half of the maximum possible connections. Within the network as a whole, actors who appear to be dominant include the universities (Aston, Birmingham), businesses and business associations (e.g. British Gas, Innovation Birmingham, Sustainability West Midlands, Midlands Environmental Business Company, Encraft), consultancies (Cenex, Encraft) and bodies linked to Birmingham City Council (West Midlands

Integrated Transport Authority, BCC Sustainability Team). Focusing on connections among the members of the influential core group there are few relatively small changes in terms of order, with a City Council Cabinet Member, community organisations (Localise West Midlands, Third Sector Assembly, Community Energy Birmingham) and supply companies (Severn Trent Water, ENGIE) scoring higher compared to the entire network. The two universities and British Gas appear to be the most well-connected and influential actors (degree and closeness centrality). However, they score slightly less high in terms of betweenness centrality, especially when considering links within the network as a whole. Here, the top position of Solihull Metropolitan Borough Council is potentially less interesting and is likely due to a lack of differentiation between different departments. Other most central actors include a consultancy company (Encraft, now Egnida) and the Greater Birmingham and Solihull Local Enterprise Partnership (GBS LEP). Their importance is reduced within the core group indicating that their key role is in connecting less influential actors to the core.

Due to the significantly higher membership of the Environmental Forum/Umweltforum than other decision-making arenas, actors that are the most well-connected in Frankfurt's network as a whole are the ones involved in the Environmental Forum/Umweltforum, and at least one other arena. These include municipal companies (Mainova, Frankfurt Airport, ABGnova), the Council's Environment Department/Umweltamt, energy co-operatives (Sonneninitiative, SolarInvest Main-Taunus), local investment bodies and banks (Burger AG fur Nachhaltiges Wirtschaften, Frankfurter Sparkasse, GLS Gemeinschaftsbank), regional public sector bodies (Landessportbund Hessen, Stadt Maintal, Regional Authority/Regionalverband, Energy Agency/Energiereferat) and a university. Within the core group, however, public sector bodies (Regional Authority/Regionalverband, Energy Agency/Energiereferat) dominate. The market sector is represented by collective professional organisations and associations with a local focus (IHK Chambers of Commerce and Industry, Handwerkskammer FRM). The high betweenness score of the Regional Authority/Regionalverband is notable, despite not being involved in the Environmental Forum/Umweltforum. In addition, betweenness centrality also indicates a potentially more influential role for energy co-operatives (specifically, Sonneninitiative) besides public sector bodies and utilities than what degree centrality suggests. Political actors appear in less central positions in Frankfurt's network compared to Birmingham's. Closeness centrality scores broadly correlate with degree centrality in Frankfurt's network too.

In Budapest, the highest scoring actors remain the same across all three centrality measures, and also between the whole network and the core group: BFVT (the municipal city planning agency), the Deputy Mayor (responsible for overseeing urban development), and the City Council's Panning and Project Management departments (tasked with the coordination and management of funding applications from Central Government and the European Regional Development Fund). In the relatively small network, opportunities for bridging activities are limited – consequently, betweenness centrality scores are comparatively low. The utilities' and Environment Team's low scores relative to the central actors indicate a lack of connection between decision-making/implementation and operational management/progress monitoring.

In summary, the actor-level analysis based on degree, betweenness and closeness centrality measures, reveals a comparatively strong position of universities and businesses in Birmingham's network as opposed to the other two networks. Most key actors operate in the region or nationally rather than at the city scale, and can access to funding streams. As such, Birmingham City Council and its dedicated Sustainability Team may find it difficult to influence these actors, and through them, the network itself. The central position of the universities suggests that energy transitions may have been driven (at the time of data collection) by research and pilots rather than large-scale transitions.

In Frankfurt, centrality scores indicate a potentially high influence of the public sector within the energy transition network, as well as a lack of integration with the market sector. Community organisations, specifically energy co-operatives appear to be in a more central position than in the other two networks.

In Budapest, the network is overwhelmingly dominated by public sector actors. The most influential actors are those higher up in the public sector hierarchy, and large departments. These central actors have no specific responsibility for sustainability or energy systems – indicating a potential lack of priority for carbon emissions reduction from the energy sector with energy transition being considered mainly in conjunction with other goals.

Furthermore, betweenness centrality scores also highlight higher levels of integration and reduced potential to take up bridging roles in Birmingham's network compared to Frankfurt's. As betweenness centrality is associated with control over flows of information and influence, this can result in coordination difficulties in Birmingham (e.g. for the City Council or its Sustainability Team).

Comparing the city cases – key findings

The reconstruction and analysis of the sustainable energy networks demonstrates that all three cities have been engaging with this agenda for some time now. Participative and collaborative decision-making arenas involved various actors in each city, potentially allowing for network forms of governance to arise. In addition to some similarities, the structural analysis of these existing energy transition networks in Birmingham, Budapest and Frankfurt also revealed significant differences among the networks. These are important considerations for the potential of networks to facilitate and support low-carbon transitions in the case study cities. First, these existing network relationships are likely to have implications for new interventions (TM-backed transition arenas or new intermediary organisations) and therefore their structural features must be considered when deciding about setting up new arenas or reorganising existing ones. Second, the structural differences identified in this chapter are likely to be caused by specific contextual factors which enable or constrain pathways for network development. As such, they also need to be taken into account when designing new network interventions or evaluating the role and potential of networks in energy transition governance in different cities.

Among the existing networks, Birmingham's and Frankfurt's appear to be more similar, whilst Budapest's network is an obvious outlier. It is significantly weaker and more exclusive than in the other two networks, evidenced by the small network size, the lack of formal participative/collaborative decision-making arenas, the lack of integration between the public sector and externals, and the apparent persistence of public sector internal hierarchies within the network. On the level of whole networks, Frankfurt's and Birmingham's are somewhat different in terms of network size, but fairly similar according to the shares of different sectors (public, market, third sector/community). However, network density (cohesion) in Birmingham's whole network is higher than in Frankfurt's. By contrast, the size of the core groups of the two networks is fairly similar, as well as their density. This represents greater separation, and potentially higher degree of internal hierarchy, between the core group and network periphery in Frankfurt compared to Birmingham. Sectoral shares remain similar in Birmingham's whole network and core group, signalling a comparatively high influence of the market sector in Birmingham. In Frankfurt, the core group is dominated by public sector bodies, including municipal departments and companies.

Similarities and differences among the networks in the three cities therefore can be summarised in terms of *integration* and *power relationships.* Integration refers to overall cohesion (density), participation from the different sectors (public, market, third/community), as well as 'permeability' – providing opportunities for new actors to join. Power relationships denote network-internal hierarchies, including sectoral dominance as well as actor centrality. According to these characteristics, Birmingham's network appears to score highest in terms of integration among network actors and among different sectors. It is the most polycentric network of the three, with little evidence for network-hierarchical relationships, allowing for more horizontal network interactions to develop among the actors. In this polycentric network, Local Authority bodies (City Council Cabinet Member, Sustainability Team) are just one of the many network actors. Frankfurt's network is somewhat different, with lower overall integration among the actors (density) but with a varied membership involving actors from all three societal sectors. However, it exhibits a higher degree of internal hierarchy between public sector bodies and network participants external to the local/regional authorities. Local Authority bodies (Energy Agency, Regional Authority, Environment Department) therefore may have more potential to influence and coordinate network processes compared to Birmingham. Integration in Budapest's network appears to be relatively low, shown most evidently by the lack of integration of external partners. Public sector internal hierarchical relationships to persist more than in the other two cities – limiting the potential for horizontal network interactions to develop. These different network structures are likely to provide space for different types of network processes, and different options for steering or coordinating these for the organisation to whom this role is assigned. As both Transition Management and the intermediation perspectives involve a form of network steering or coordination (transition manager, intermediary organisation), this has implications for the application of such frameworks in different cities.

These structural differences are attributed to the existence of contextual factors that have an impact on the formulation and operation of energy transition networks in Birmingham, Frankfurt and Budapest. Such local contextual characteristics are thus likely to act as constraining and enabling factors in terms of the role of network forms of governance in energy transitions in the different cities. The following three chapters investigate how the local 'context' has been influencing the emerging network forms of governance in Birmingham, Frankfurt and Budapest.

Notes

1 Free to download from www.gephi.org.
2 Available from gephi.org/plugins/#/plugin/multimode.
3 Available from https://github.com/tnochta/Networks-energy-transitions-European-cities under Creative Commons licence for reuse with attribution (CC BY 4.0).
4 A complete list of network visualisations (with node sizes based on degree, betweenness and closeness centrality scores) are available from https://github.com/tnochta/Networks-energy-transitions-European-cities under Creative Commons licence for reuse with attribution (CC BY 4.0).
5 Available from https://github.com/tnochta/Networks-energy-transitions-European-cities under Creative Commons licence for reuse with attribution (CC BY 4.0).

References

Borgatti, S. P. and Halgin, D. S. (2011) 'On network theory', *Organization Science*, 22(5), pp. 1168–1181. doi:10.1287/orsc.1100.0641.

Bryman, A. (2016) *Social Research Methods*, 5th edn. Oxford: Oxford University Press.

Burt, R. S. (2017) 'Structural holes versus network closure as social capital', in R. Dubos (ed.), *Social Capital: Theory and Research*, pp. 31–56. Taylor & Francis. doi:10.4324/9781315129457-2.

Climate-KIC (no date) Transition Cities. Available at: www.climate-kic.org/projects/transition-cities/ (Accessed: 27 February 2020).

Copus, C., Roberts, M. and Wall, R. (2017) *Local Government in England: Centralisation, Autonomy and Control*. Palgrave Macmillan. doi:10.1057/978-1-137-26418-3.

Freeman, L. C., Roeder, D. and Mulholland, R. R. (1979) 'Centrality in social networks: ii. Experimental results', *Social Networks*, 2(2): pp. 119–141. doi:10.1016/0378-8733(79)90002-9.

Grandjean, M. (2015) GEPHI – Introduction to network analysis and visualization. Available at: www.martingrandjean.ch/gephi-introduction/ (Accessed: 29 February 2020).

IEA (2017) Cities lead the way on clean and decentralized energy solutions. *IEA*. Available at: www.iea.org/news/cities-lead-the-way-on-clean-and-decentralized-energy-solutions (Accessed: 29 February 2020).

Kuhlmann, S. and Wollmann, H. (2014) *Introduction to Comparative Public Administration: Administrative Systems and Reform in Europe*. Edward Elgar Publishing.

Marcussen, M. and Olsen, H. P. (2007) 'Transcending analytical cliquishness with second-generation governance network analysis', in J. Torfing and M. Marcussen (eds), *Democratic Network Governance in Europe*, pp. 273–292. Palgrave Macmillandoi:10.1057/9780230596283_144.

Marsden, P. V. (2015) 'Network Centrality, Measures of', in N.J. Smelser and P.B. Baltes (eds), *International Encyclopedia of the Social & Behavioral*

Sciences, 2nd edn, pp. 532–539. Elsevier. doi:10.1016/B978-0-08-097086-8.43115-6.

Nochta, T. and Skelcher, C. (2020) 'Network governance in low-carbon energy transitions in European cities: A comparative analysis', *Energy Policy*, 138. doi:10.1016/j.enpol.2020.1112988.

Prell, C. (2012) *Social Network Analysis: History, Theory and Methodology.* SAGE.

Rockström, J.*et al.* (2017) 'A roadmap for rapid decarbonization', *Science*, 355 (6331), pp. 1269–1271. doi:10.1126/science.aah3443.

Swianiewicz, P. (2014) 'An empirical typology of local government systems in eastern Europe', *Local Government Studies*, 40(2), pp. 292–311. doi:10.1080/03003930.2013.807807.

Transition Cities (2015) Maximising Europe's low carbon activities: Moving from individual projects to challenge-led, transition programmes. EIIT. https://eit.europa.eu/library/maximising-europe%E2%80%99s-low-carbon-activities-moving-individual-projects-challenge.

3 Case study I: Energy transitions in Birmingham

Introduction

The analysis of the organisational networks of Birmingham, Budapest and Frankfurt indicated existing engagement with the sustainability and low-carbon energy agendas in all three cities. The processes of developing and implementing decarbonisation plans involved some form of interaction among multiple societal actors in each city. The network analyses presented in the previous chapter aimed at capturing these emerging collaboration patterns among the actors, and highlighted significant structural differences between the networks.

The investigation into participative and collaborative decision-making initiatives involving multiple actors uncovered a relatively large and well-connected network in Birmingham. Of the three cities, Birmingham's network showed the highest levels of integration among network actors and across different sectors (public, market, third sector and academia/research; see Figure 3.1). Power relationships within the network appeared to be fairly balanced, between actors and between the public sector and the rest of the network, resulting in a polycentric network structure. The public sector was mainly represented by the Cabinet Member for Sustainability and the City Council's Sustainability Team rather than traditional large municipal departments. This suggests that the issues of sustainability and low-carbon transitions are potentially not well-integrated into the wider operations of Birmingham City Council. In addition to public sector bodies, other central actors in the network included supply companies and other market sector actors, universities (Aston, Birmingham) and to some extent, community organisations.

This chapter seeks to link the structural network characteristics (the degree of integration among network actors; network-internal power relationships) to the local 'context' in which the network has been

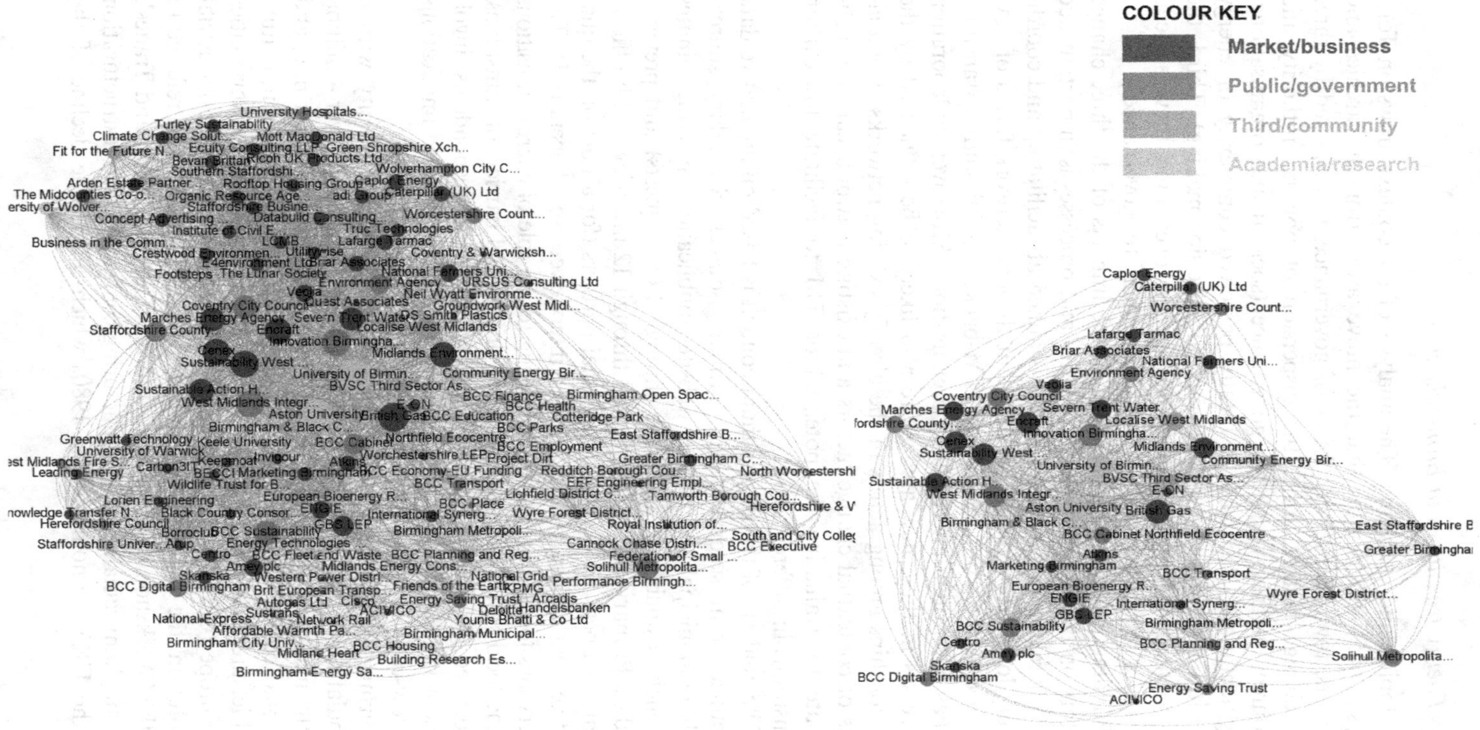

Figure 3.1 The energy transition network in Birmingham – whole network and core group.

developing and operating. The analysis of the context draws on Torfing's (2005) observation that the move away from hierarchies and markets, towards network forms of governance, is facilitated by three key change factors. First, the dispersion of powers, authority and resources in energy systems governance (societal and organisational fragmentation). Second, the locally perceived complexity of energy transitions (complexity in contemporary decision-making). Third, the development of the sustainability agenda over time (new societal dynamics related to a shift from 'government' to 'governance').

Building a better understanding of the impacts of the three change factors on the local network structures and processes is important due to their relevance for future attempts aimed at facilitating and coordinating low-carbon urban development through network forms of governance – for example in new TM-backed collaborative transition arenas or through intermediary organisations. Moreover, collaboration processes in these new arenas are also likely to be influenced by the relationships among actors in the already existing networks. The analysis of contextual change factors and existing networks can thus contribute to evaluating the potential of TM to facilitate low-carbon transitions in different cities.

The analysis of the Birmingham context is based on qualitative data obtained from secondary and primary sources. Secondary sources include municipal and partner websites, publications, meeting minutes (publicly available or accessible through FOI requests) and previous studies. Primary data was collected through 12 interviews with officers from the Local Authority and external partners (for a note on the qualitative data collection and a list of interviewees see Appendix I). The interviews focused on questions relating to the role of energy transitions in the local sustainability agenda, the drivers and barriers of transitions, associated governance initiatives and processes, the respondents' involvement in these and the experiences gained, and perceived implications for low-carbon energy transitions in the city.

Birmingham is the United Kingdom's second largest city with a population of over 1.1 million within the city limits, and over 4 million in the metropolitan area. The city is a major transportation hub, and regional centre for employment, business, industry, education, culture and leisure. It is among the youngest and most ethnically diverse cities in Europe (GBCC, 2019). Situated in the heartland of the industrial revolution, the city has a rich industrial history. Historically referred to as the 'Workshop of the World' and the 'City of a Thousand Trades', Birmingham remained a manufacturing centre until the restructuration of the UK's economy in the 1980s which led to the decline of the manufacturing sector. At the height of the crisis, unemployment levels in the city climbed as high as 25% (BCC, 2008b).

The past decades have thus been characterised by an aspiration to reinvent the city and its economy, and to bring back the prosperity and world-leading role in innovation for which Birmingham has once been famous (BCC, 2002). From the 1990s, large-scale regeneration projects started which aimed at generating growth in the conference and hospitality sectors. Subsequent retail investment contributed to a move away from the previous industry-based local economy to services and finance. The decline of industrial production and its effects inspired a local policy agenda around making Birmingham 'cleaner, greener and safer' (BCC, 2002). In the following decades, a plethora of policies and interventions have been developed to facilitate sustainable urban development.

Energy systems governance in Birmingham and England

Patterns of organisational and societal fragmentation in the governance of energy systems describe the ways which powers and responsibilities are dispersed between different levels of government, and the public, private and civil societal domains, in the energy sector in the United Kingdom. This requires investigating how (1) energy policy and regulations are made and by whom, (2) the organisational setup of the ownership and maintenance of energy infrastructures (electricity, natural gas and other decentralised systems), as well as (3) the ways in which energy supply is organised in the English context, and specifically in Birmingham.

First, energy policy-making and regulation powers and responsibilities are retained by Central Government bodies. Energy policy falls within the remit of the Department for Business, Energy & Industrial Strategy (BEIS). Regulation is undertaken by the 'Gas and Electricity Markets Authority' (GEMA), which operates through the Office of Gas and Electricity Markets (Ofgem) and the Secretary of State for Business, Energy and Industrial Strategy. Ofgem is a non-ministerial department within the UK Government, responsible for overseeing the everyday functioning of the national energy markets and for the protection of consumer rights by promoting and safeguarding market competition (Hassan and Majumder-Russell, 2014).

Second, the structure of electricity and gas systems are highly centralised in the United Kingdom. The national systems consist of high-voltage electric and high-pressure gas transmission lines and regional distribution networks. Transmission systems are operated by National

Grid Plc as a natural monopoly. The electricity distribution networks are operated by 14 licensed distribution network operators (DNOs) whose ownerships belong to six companies. There are eight regional distribution systems for natural gas, four of which are owned by National Grid, including Birmingham's West Midlands network. The region's electricity network is operated by Western Power Distribution Plc, owned by a US corporation registered in Philadelphia (PPL Corporation). Neither the transmission nor the distribution system operators are allowed to participate in energy supply, i.e. buy or sell electricity or natural gas.

Third, energy supply markets are liberalised and operate on the basis of market competition between private energy producers and suppliers, overseen by Ofgem. Trade is organised within the frames of competitive wholesale and retail markets, dominated by six large energy supply companies (British Gas, EDF Energy, E-ON, npower, Scottish Power and SSE – Scottish and Southern Energy). Many of these companies are international corporations or are owned by one, with headquarters either in the UK (e.g. British Gas) or abroad (EDF Energy, E-ON, npower). Despite continuous decrease in recent years, the total market share of the 'Big Six' is still above 70% both in electricity and gas supply (Ofgem, 2020).

English Local Authorities, including Birmingham's, have not been able to play any role in energy policy, operation or supply since the nationalisation of energy systems in the 1940s. In fact, until recently they were barred from obtaining energy supply licences and setting up or operating municipal energy companies (Webb, Hawkey and Tingey, 2016). The first Local Authority-owned, not-for-profit energy company in the UK – Robin Hood Energy – only started operating in 2015 in Nottingham (robinhoodenergy.co.uk). The company, however, has so far failed to become profitable in an energy market that favours large-scale production and centralised supply (Bolton and Foxon, 2013), and more recently has been threatened with losing its licence over disputes with Ofgem (Ofgem, 2019).

In Birmingham, distributed generation projects have been carried out by private sector partners, such as Veolia (waste incineration plant) and Engie (CHP and district heating). In addition to legal barriers, the lack of experience, and financial and human resources at the local level also contributes to maintaining the local authority's marginal position in the governance of local energy systems. Furthermore, the centralised nature of energy production, distribution and supply, and the resulting grid architecture optimised for one-way flows of electricity and gas, acts as a barrier for small (local/regional) companies, community energy schemes (e.g. Community Energy Birmingham) and producer-consumers to join the market and scale their activities.

Complexity and uncertainties of sustainable energy transitions

The decarbonisation agenda has a particularly strong focus on innovation and 'green growth' in the United Kingdom (Wanner, 2015). In tandem, the political will and support for the low-carbon policy and implementation remains fairly superficial and politicised. In the energy sector specifically, the frequent organisational restructuring in terms of roles and authority both within the public sector and between the public and private sectors, has also been contributing to a sense of lack of leadership for low-carbon development. As a result, policy commitments (achieving net-zero emissions by 2050) remain at odds with implementation and delivery (CCC, 2019).

This situation is further nuanced by the centralised energy systems architecture and liberalised private-sector-dominated energy supply. There are high investment requirements and costs associated with infrastructure upgrades to accommodate small(er)-scale production, carbon-neutral technologies and two-way energy flows. Consequently, the economic viability of implementing low-carbon plans is often questioned by the private sector.

Besides public or market sector leadership, societal pressure could also help pushing investment into and implementation of decentralisation efforts and restructuring of energy systems towards more distributed production and distribution. However, the historical accessibility of fossil fuel reserves has meant that a widespread concern over the security of supply has not (yet) developed in the UK. The aspiration for energy autonomy is therefore not present as a driver for moving towards more small-scale, renewable production.

In addition to the lack of market (no strong business case for decentralisation which could facilitate more small-scale renewable generation) and social (no widespread concern over energy security) pressures, progress in the public sector (Central Government) with developing detailed energy transition strategies and pathways has also been slow. Consequently, despite a broad consensus on long-term carbon emissions reduction goals, the expected contribution of the energy sector including the role of decentralised technologies has remained unclear. Decentralisation has mainly been considered through ideological lenses (Interview 1.05). This is problematic in the UK political context which has traditionally been dominated by regular shifts between the Conservative and Labour Parties, with largely opposing ideologies and perspectives on social and economic frameworks, as well as regarding the 'appropriate' roles for the public and private sectors in energy supply.

The rationale of energy transition in Birmingham is also affected by the aforementioned issues. The 'green growth' narrative is dominant locally, similarly to the national level. Decarbonisation is considered as part of a broader low-carbon innovation agenda aimed at delivering economic growth, and reinstating Birmingham's status as a place of innovation, while also reducing carbon emissions. The objective set out by the City Council's low-carbon agenda is to 'make Birmingham more prosperous, healthier, fairer, resource-efficient and better for business' and in doing so, enhance citizen's quality of life and well-being (BGC, 2013a, p. 3). This emphasis on green economy has been further strengthened by the Local Enterprise Partnerships' role in distributing European and national development funds (GBS LEP, 2016).

The lack of administrative leadership and political consensus are characteristic features of Birmingham City Council's approach to low-carbon transitions. Commitments are relatively superficial – for example, despite having committed to 60% carbon emissions reduction by 2027, and developed a 'Roadmap' to achieve this target, the expected role of the energy sector in the overall reduction remains vaguely defined (BGC, 2013b). The Council had appointed a Cabinet Member for Sustainability – renamed as Cabinet Member for Smart, Green and Sustainable City in 2018. However, progress towards a more coordinated and integrated approach to delivering the carbon emissions reduction targets across different Council departments, and between the elected and the executive branches, has been slow. The lack of support from senior officers from traditional large Council departments has been cited by key informants as a potential barrier to developing such an integrated approach (Interview 1.05; Interview 1.12).

Furthermore, due the recent cuts to budget and personnel in English local governments, and the historical hierarchical-dependent relationship between the central and local levels, there is a dominant view that prioritises the delivery of core services as opposed to broadening the portfolio of tasks, for example with energy infrastructure management (Interview 1.12; Interview 1.03). Consequently, Birmingham City Council's Sustainability Team needs to rely on external stakeholders and experts to justify decarbonisation strategies and goals towards other departments within the Council.

In the UK energy systems context, the available external partners are likely to be large, often multinational, incumbent corporations. The lack of legal, regulatory or operational responsibilities at the local level contributes to imbalance in terms of powers and resources between the local authority (and the Sustainability Team specifically) and their external partners (Interview 1.04). Moving the locus of decision-making away from the local level is likely to contribute to path dependence in the reorganisation of energy supply and grid operation for a variety of reasons. First, local actors and new players remain marginalised in decision-making processes related to local low-carbon development, with restricted options available to join the energy sector. Second, the profit-sensitivity of key players in the UK energy supply (international corporations), coupled with considerable lobby powers towards Central Government, are likely to act as barriers to transformative change. Third, engaging with Local Authorities is not a priority for key energy systems stakeholders.

Despite the high degree of uncertainty regarding future net-zero carbon energy systems, Birmingham City Council established a target which exceeds national (interim) commitments by aiming to reduce carbon emissions by 60% by 2027 (BGC, 2013a). According to the Carbon Plan Analysis commissioned by the Council the national efforts will only result in a 40% reduction, with local action necessary to cover the remaining 20% (BGC, 2013c). While key focus areas have been identified by the Birmingham Carbon Roadmap (BGC, 2013b) – Buildings & Efficiency, Energy & Resources, Transport & Mobility, Natural Capital & Adaptation, Catalysing Green Growth & Behaviour – specific targets for each thematic area are lacking. As such the Roadmap is considered rather as a collection of ideas than a coherent strategic plan which assigns roles and responsibilities and ensures progress (Interview 1.09). In addition, the Green Commission of Birmingham City Council, which has been set up to deliver the low-carbon ambitions, has been described as an advisory board without a clear responsibility or authority and a 'bit of a talkshop' where 'lots of good ideas are being talked about, but implementation is a shortcoming' (Interview 1.02).

The historical development of Birmingham's sustainable energy agenda

The previous sections discuss how the broader political-organisational structure and culture in the United Kingdom influences energy transitions in Birmingham. This section takes a historical perspective and introduces the development of the local sustainability agenda and the associated societal dynamics and changes in the governance of sustainability and low-carbon energy.

Whilst Birmingham City Council reacted quickly to international developments in the sustainability domain in the early 1990s (by developing a Local Agenda 21; (BCC, 2002)), commitments and priorities have been subject to frequent change ever since, mostly following political changes, including changes in the leadership of the Council between the Labour Party and coalitions of the Conservative and the Liberal Democrat Parties (the third most important political party in England). Figure 3.2 shows an overview of the different governance initiatives with a relevance to the development of the local low-carbon agenda.

A 'Local Agenda 21 Team' was set up within the City Council in the beginning of the 1990s in response to the call by the UNFCCC for developing local agendas for sustainable development. The Team started engagement with local stakeholders and citizens via the Birmingham Sustainability Forums which provided opportunity for regular consultations on priorities for local action to develop a Birmingham Local Agenda 21 (LA 21), published in 1997 (BCC, 2002). Based on the LA 21, various sustainability strategies and action plans were produced in the following years focusing on key themes of social issues (education, healthcare and access to green spaces), waste management, improving safety across the city and making local governance more transparent, democratic and participative (BCC, 2002).

The Local Agenda 21 Team was abolished in 2003/2004 by the incoming Conservative–Liberal Democrat leadership (City Council) and public sector responsibility and ownership of the sustainable development issue was transferred to the regional level. This was achieved through the creation of a Sustainability Department as part of the newly established Regional Development Agency, Advantage West Midlands. The Regional Development Agencies were non-departmental public bodies charged with developing five to ten years Regional Economic Strategies in their regions, coordinated by the central government (controlled at the time by the Labour Party) (LGA, 2010). Sustainability West Midlands (SWM) has been created as a regional forum for collaboration and deliberation in the sustainability domain which brings together public and private sector actors and community organisations. SWM is still in operation today, despite the abolishment of the Regional Development Agencies in 2012 by the Conservative – Liberal Democrat Coalition Government (national level).

On the city level, continued debate over the sustainability agenda was secured via establishing a City Strategic Partnership, supervised by Advantage West Midlands and Birmingham City Council. The

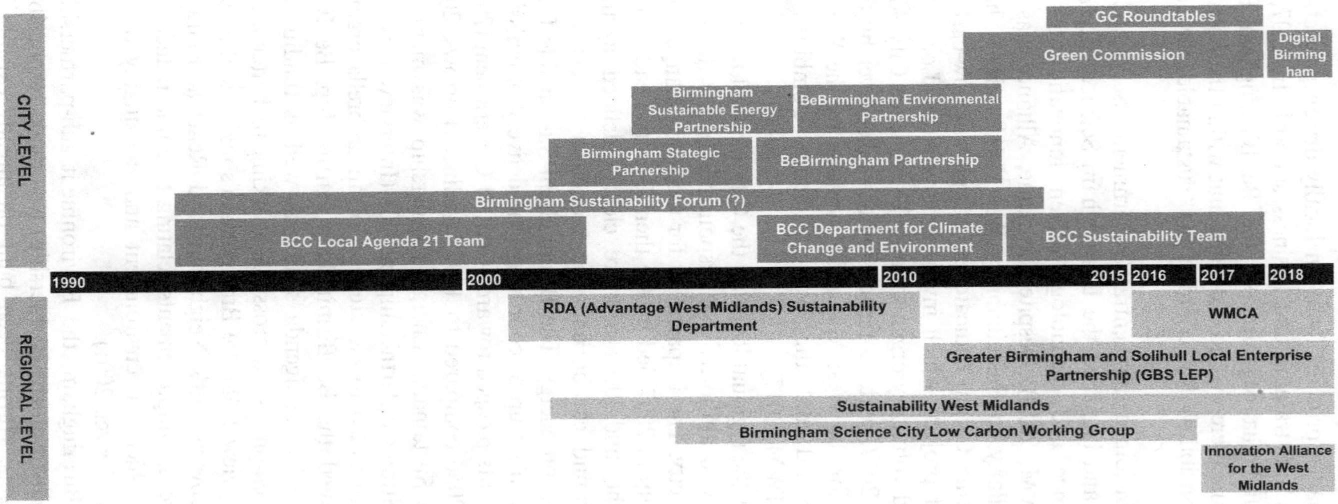

Figure 3.2 Overview of the development of low-carbon energy governance initiatives in Birmingham.

Partnership had a primary focus on locally driven projects for regeneration, skills and training and business support. In 2007, it became part of an umbrella organisation, the Be Birmingham Partnership. During this time there was no dedicated unit within the Local Authority with a responsibility for coordinating sustainable development in Birmingham.

Birmingham joined the Central Government's Science Cities initiative in 2005, and launched the Birmingham Science City Alliance to promote science and technology driven innovation for economic growth (IAfWM, 2018). Despite the name, Alliance was a regional initiative similarly to Sustainability West Midlands. It had a Low Carbon Working Group dedicated to supporting low-carbon innovation and pilot projects which involves universities, Local Authorities, businesses and voluntary organisations. The Science City Alliance was shut down in 2017, with its responsibilities taken over by the Innovation Alliance for the West Midlands, overseen by the West Midlands Innovation Board and the West Midlands Combined Authority (WMCA) (IAfWM, 2017).

Successes at the regional level led the City Council to re-establish an internal unit with authority and responsibilities for the sustainability agenda of the city. The Department for Climate Change and Environment was set up in 2007, led by the then Deputy Leader of the Council. During the mid-2000s a more clearly defined agenda started developing around carbon emission reduction targets and to some extent, low-carbon energy. The Birmingham Sustainable Energy Partnership has been set up to coordinate the delivery of emissions reduction commitments pledged towards Central Government (20% by 2010 and 80% by 2050 compared to 1990 baseline; (Burrows, 2004)). Subsequently, the Sustainable Energy Partnership was merged into the Birmingham Strategic Partnership in 2008 (Interview 1.08).

Further mergers led to the creation of an umbrella organisation in 2007/2008 named the Be Birmingham Partnership. Be Birmingham became responsible for agenda setting as well as funding allocation, with a strong emphasis on cross-silo working to better address local problems (Interview 1.05). Be Birmingham's main funding source was the Central Government's Neighbourhood Renewal Fund. This was linked to delivering improvements relating to crime reduction, employment, health, education, environment and the quality of housing in deprived areas (Weaver, 2001).

Within Be Birmingham, the Environment sub-partnership (BEP – Birmingham Environment Partnership) was made responsible for sustainability targets. Having been built on already existing initiatives,

BEP aimed to make the city 'cleaner and greener' through facilitating collaboration across community groups, businesses and academia. Drawing on Neighbourhood Renewal Funding, BEP's budget was supporting small-scale, community-led projects (Interview 1.05). From 2008, the Sustainable Energy Partnership was integrated into the Birmingham Environment Partnership, shifting its focus towards delivering carbon emissions reduction targets set out by the Local Area Agreement of Birmingham. Strategic action to achieve these targets regarded growth in low-carbon economy, increasing resource efficiency and reducing waste, preparing climate change adaptation strategies promoting infrastructure development (energy, water and wastewater), and protecting biodiversity and green spaces (Be Birmingham, 2010).

One of the main drivers of the change was the 2008 Climate Change Act approved by the Central Government. The Climate Change Act is the UK's main legislation for reducing carbon emission, aiming for net-zero emissions (then 80%, since increased to 100% reduction) by 2050 (HM Government, 2008). In 2007/2008 Birmingham City Council also committed to a 60% carbon emissions reduction target by 2026, and established a Department for Climate Change and the Environment – both championed by the then Deputy Leader of the Council (Be Birmingham, 2010). In 2009, the City joined the European Covenant of Mayors initiative and devised an Energy Action Plan 2005–2020 aimed providing a general direction for energy transitions rather than a detailed implementation plan (BCC, 2009).

In 2010, the Central Government was taken over by a Conservative–Liberal Democrat coalition, whilst Labour regained leadership of Birmingham City Council. The Regional Development Agencies, funding streams associated with them (e.g. Neighbourhood Renewal Fund) were closed and abolished by 2012 (Broughton, Berkeley and Jarvis, 2011). These changes led Birmingham City Council to revise targets and action plans and reconfigure organisational structures for low-carbon development. Be Birmingham was abolished, and the responsibilities of the BEP were transferred to a more strategic level by setting up a new Green Commission led by the Cabinet Member for Sustainability, and coordinated by the Sustainability Team. The Sustainability Team was a successor to the Department for Climate Change and Environment. The Green Commission (at the time of data collection still in operation) was a strategic advisory body for Birmingham City Council on issues related to sustainability. Despite the strategic role, the Commission had no direct funding to support its activities.

The restructuration led to some frustration among the City Council's external partners, particularly community organisations embedded in Be Birmingham and involved in the delivery of small-scale local projects:

> A lot of the people that I used to deal with are gone, the people that are left have more of a regional focus, which looks good on paper, but I don't think it actually means … I don't think it does anything positive in terms of local delivery … [T]ime after time you see stuff that is either not advertised, you see bits of work being commissioned by Birmingham City Council that … have been given to people outside of Birmingham when there are people in Birmingham that could do it.
>
> (Interview, 1.08)

In its first year, the Green Commission realigned the city's carbon emissions reduction target with the national carbon budgets set out by the 2008 Climate Change Act. The new target, which is still in place today, became 60% reduction by 2027 (BGC, 2013a). The result of a scoping exercise aimed at identifying core themes, activities and pathways towards achieving the target were published as Birmingham's Low Carbon Roadmap (BGC, 2013b). The core areas of action identified in the Roadmap led to setting up five thematic working groups ('Roundtables') as part of the Commission: Green Growth, Buildings & Efficiency, Energy & Resources, Transport & Mobility and Natural Capital & Adaptation. The Roundtables were tasked with implementing the low-carbon agenda published in the Roadmap.

On the regional level, the Central Government established the Local Enterprise Partnerships' (LEPs) in 2011 to take up some of the roles of the Regional Development Agencies. LEPs are aimed at facilitating collaboration between Local Authorities and businesses with a responsibility for deciding about priorities for investment in assets such as roads, buildings and facilities and to develop 'Enterprise Zones' (HM Government, 2012). Enterprise Zones are areas dedicated for regeneration to deliver economic growth. In Birmingham, the Greater Birmingham and Solihull LEP was tasked with overseeing the development of the Birmingham City Centre Enterprise Zone (GBS LEP, 2016). Having access to significant funding streams (e.g. European Regional Development Funds), the Greater Birmingham and Solihull LEP quickly became influential in the Green Commission as well. As a regional body (with representation from nine Local Authorities), the involvement of LEP members in the Green Commission once again started shifting the emphasis towards a regional approach rather than focusing on local issues and projects in Birmingham.

In 2016 another regional body was set up through a 'devolution deal' opportunity offered by the Central Government, to fill the gap created by the abolishment of the Regional Development Agencies. Combined authorities are legal structures formed by two or more collaborating Local Authorities (HM Government, 2016). Birmingham became part of the West Midlands Combined Authority which has 19 Local Authority members and brings together 3 LEP areas. In continuation of the renewed 'regionalisation' of the low-carbon agenda, the Sustainability Team within the City Council, as well as the Green Commission, were abolished in 2017/2018. The sustainability agenda has become intertwined with the smart and green city visions within the Council, championed by the Cabinet Member for Smart, Green and Sustainable City.

The West Midlands Combined Authority (WMCA) Board declared climate emergency in 2019, and set a net-zero carbon target for the region by 2041, with an aspiration to support the decarbonisation action with an investment equivalent to approximately 1–2% of the regional gross value added (WMCA, 2019). In conjunction with the review of emissions reduction targets, the WMCA also established Energy Capital, a 'smart energy innovation partnership for the West Midlands' (www.energycapital.org.uk) with a role of coordinating collaborative public-private investment for low-carbon energy projects.

Outcomes: Decentralised energy in Birmingham

Currently, only 2% of the Birmingham's electricity and heat demand is generated locally (Lee, Quinn and Rogers, 2016), mainly produced by a waste-to-energy plant and combined heat and power plant (CHP) feeding a city centre district heating network. The waste-to-energy plant was built by Veolia in 1996. Veolia is in a contractual relationship with the City Council which is currently set to conclude in 2024. The City Centre District Energy Scheme is operated by the Birmingham District Energy Company, owned by Engie. The network, however, is limited in extent compared to other European cities.

Further low-carbon energy installations include a sewage treatment plant in Minworth (owned by Severn Trent Water) the Tyseley Energy Park (coordinated by Webster and Horsfall) and a number of solar power installations. The Minworth plant uses an innovative thermal hydrolysis technology to produce biogas (Severn Trent Water, 2017). The Tyseley Energy Park (Energy Capital, no date) is set to become one of the Energy Innovation Zones supported by Energy Capital. One example of solar power installations is the Alexander Stadium's solar roof completed in 2004/2005 (BCC, 2008a). Solar power has also been championed by Community Energy Birmingham, a local renewable energy co-operative, who installed seven community energy schemes to date (CEB, no date).

While Birmingham City Council supports these low-carbon energy projects, it has not been directly involved in setting up or managing them. The largest energy saving initiative to date has been the Birmingham Energy Savers (set up in 2006) which aimed at contributing to tackling fuel poverty. The initiative was led by the Council, and supported by Central Government funding through its Green Deal, in partnership with an external stakeholder: the energy company Carillion. Carillion was a major contractor to the UK public sector which collapsed in 2018 (BBC, 2018). Birmingham Energy Savers was designed to provide financial and professional assistance to carry out energy-efficient retrofit of existing buildings, with a high potential to contribute to emissions reduction due to the poor quality of Birmingham's housing stock (SWM, 2015). However, due to political changes and the discontinuation of the Green Deal, the project did not deliver its initial aim of retrofitting 60,000 homes deemed inefficient (E to G) according to the European Energy Performance Certificate (EPC) rating scheme (BPIE, 2014).

In addition to the 'on the ground' initiatives discussed above, considerable research activity has been undertaken at major universities of the city, including the University of Birmingham, Aston University and Birmingham City University. The research and pilots mainly focus on distributed (thermal) energy storage and bioenergy.

This overview of the distributed energy generation projects demonstrates that Birmingham City Council, and the local and regional public sectors in general, have not had a significant or direct role in setting up and managing energy infrastructures. In addition to periodic shifts in the City Council's interest in leading the low-carbon agenda, it was found that experience in implementation may also be lacking within the local authority. Projects are mostly operated by the private sector, often by international corporations, and the opportunities for local actors to enter the energy sector are limited.

Networks, context and energy transitions in Birmingham

In the United Kingdom, energy systems are dominated by Central Government bodies and national and international corporations (Devine-Wright and Wiersma, 2013). In spite of recent developments, including a surge in new small energy supply companies joining the market, the move towards decentralisation in energy production and distribution has been relatively slow. The historical exclusion of Local Authorities (and other local actors) from energy systems governance has had a negative effect on local capacity and capability to manage energy infrastructures, with implications for low-carbon energy transitions in the city.

Low-carbon energy transitions are characterised by high levels of uncertainty and complexity both at the national and the local levels. Despite ambitions carbon emissions reduction targets, implementation and delivery are lagging behind. The structure of the physical energy grids, the organisational landscape of energy governance, as well as the broader ideological context, act as barriers to decentralisation. In order to build capacity for energy transitions, Birmingham City Council, and its Sustainability Team, need to rely on external partners in the delivery of low-carbon initiatives. This increases the complexity of local low-carbon development not only due to the differences in terms of the interests of various stakeholders, but also due to the relatively weak powers and authority of local actors (the Local Authority among them) to coordinate implementation and delivery. In addition, it also contributes to a sense of 'inability to act' without support from the private sector. The pace of 'real-world' progress therefore is not aligned with the ambitions on paper, and energy transitions are dominated by less innovative market sector driven initiatives (district heating, waste incineration) and research and pilot projects coordinated by academic institutions (Aston University, University of Birmingham).

Moreover, the targets, action plans and organisational structures have been frequently changing in Birmingham. Corresponding with the cyclical localisation and regionalisation of low-carbon authority and responsibilities, the presence of the City Council in the governance of sustainability has also been subject to periodic shifts. In parallel with this periodic interest from Birmingham City Council to lead low-carbon development, initiatives involving mainly private sector actors remained operational. The resulting integration and stronger networks across private stakeholders, in which the local authority occupies a relatively peripheric position, strengthens the green growth agenda prioritising on economic development over low-carbon targets.

The structural network analysis presented in the previous chapter found that Birmingham's energy transition network was the most polycentric, characterised by high levels of integration and a lack of network-internal hierarchical relationships. This was linked to a relatively strong position of the market sector within the network, and a weaker position of Local Authority bodies. The contextual factors discussed above provide explanation for these network characteristics. A variety of factors weaken the position of Birmingham City Council relative to the rest of the network. These include first, the historical exclusion of Local Authorities and other local-level actors, and the strong presence of large companies, in energy systems governance. Second, the dominant ideological context characterised by lack of focus and direction to facilitate consensus, and competing and often conflicting interests (for example between growth and sustainability), result in a superficial commitment to decarbonisation and reduce the credibility of the City Council's low-carbon ambitions. Third, the periodically changing interest and intermittent presence of Birmingham City Council (and the Sustainability Team) in the network.

Whilst these factors reduce the local authority's potential to assume a leading role in energy transitions, they also contribute to a dominant view that delivering low-carbon ambitions requires support from, and collaboration with, various societal actors. Integration is further enhanced by Birmingham City Council's longstanding experience with participative forms of governance, both in relation to the broader sustainability agenda as well as other policy domains. In addition, the lack of internal consensus over low-carbon ambitions among the different Council departments drives engagement with external stakeholders to seek support for developing a more unified position within the City Council.

References

BBC (2018) Carillion collapse to cost taxpayers £148m. Available at: www.bbc.co.uk/news/business-44383224 (Accessed: 29 February 2020).

BCC (2002) Towards a sustainable city. Birmingham City Council achievements. Local Agenda 21 1992–2002. Available on request from Birmingham City Council.

BCC (2008a) Birmingham's transformation & future prospects. Available at: www.birmingham.gov.uk/download/downloads/id/3064/birminghams_transformation_and_future_prospects.pdf.

BCC (2008b) Birmingham's transformation and future prospects. Available at: www.birmingham.gov.uk/downloads/file/3064/birminghams_transformation_and_future_prospects.

BCC (2009) Birmingham Sustainable Energy Action Plan. Available at: www.special-eu.org/assets/uploads/SEAP_Birmingham.pdf.

Be Birmingham (2010) Birmingham Climate Change Action Plan 2010+. Available at: www.birmingham.gov.uk/download/downloads/id/1889/climate_change_action_plan_2010.pdf.

BGC (2013a) Birmingham's Green Commission. Building a green city. Vision statement. Available on request from Birmingham City Council.

BGC (2013b) Birmingham's Green Commission. Carbon roadmap. Available on request from Birmingham City Council.

BGC (2013c) Birmingham Carbon Plan analysis. Available on request from Birmingham City Council.

Bolton, R. and Foxon, T. J. (2013) 'Urban infrastructure dynamics: Market regulation and the shaping of district energy in UK cities', *Environment and Planning A: Economy and Space*, 45(9), pp. 2194–2211. doi:10.1068/a45575.

BPIE (2014) Energy performance certificates across the EU. BPIE – Buildings Performance Institute Europe. Available at: http://bpie.eu/publication/energy-performance-certificates-across-the-eu/ (Accessed: 29 February 2020).

Broughton, K., Berkeley, N. and Jarvis, D. (2011) 'Where next for neighbourhood regeneration in England?', *Local Economy: The Journal of the Local Economy Policy Unit*, 26(2), pp. 82–94. doi:10.1177/0269094210397427.

Burrows, P. (2004) An energy partnership for Birmingham. Birmingham Friends of the Earth. Available at: www.birminghamfoe.org.uk/what-we-do/issues-we-work-on/energy-and-climate-change/an-energy-partnership-for-birmingham/ (Accessed: 29 February 2020).

CCC (2019) Reducing UK emissions – 2019 progress report to Parliament. Committee on Climate Change. Available at: www.theccc.org.uk/publication/reducing-uk-emissions-2019-progress-report-to-parliament/ (Accessed: 29 February 2020).

CEB (no date) Our renewable energy – Community energy Birmingham. Available at: www.communityenergybirmingham.coop/about-us/what-have-we-done-so-far/ (Accessed: 24 February 2020).

Climate-KIC (no date) Transition Cities. Available at: www.climate-kic.org/projects/transition-cities/ (Accessed: 27 February 2020).

Devine-Wright, P. and Wiersma, B. (2013) 'Opening up the "local" to analysis: exploring the spatiality of UK urban decentralised energy initiatives', *Local Environment*, 18(10), pp. 1099–1116. doi:10.1080/13549839.2012.754742.

Energy Capital (no date) Tyseley Energy Park. Available at: www.energycapital.org.uk/tyseley-energy-park/ (Accessed: 24 February 2020).

GBCC (2019) Birmingham Economic Review. Available at: www.greaterbirminghamchambers.com/research-campaigning/birmingham-economic-review/ (Accessed: 29 February 2020).

GBS LEP (2016) Strategic economic plan 2016–30 – GBS LEP. Birmingham. Available at: https://gbslep.co.uk/resources/reports/strategic-economic-plan-2016-30 (Accessed: 29 February 2020).

Hassan, M. and Majumder-Russell, D. (2014) Electricity regulation in the UK: Overview. Available at: https://uk.practicallaw.thomsonreuters.com/1-523-9996 (Accessed: 29 February 2020).

HM Government (2008) *Climate Change Act 2008*. Available at: www.legislation.gov.uk/ukpga/2008/27/contents (Accessed: 29 February 2020).

HM Government (2012) 2010 to 2015 government policy: Local Enterprise Partnerships (LEPs) and enterprise zones. GOV.UK. Available at: www.gov.

uk/government/publications/2010-to-2015-government-policy-local-enterprise-partnerships-leps-and-enterprise-zones (Accessed: 29 February 2020).

HM Government (2016) *Cities and Local Government Devolution Act 2016*. Queen's Printer of Acts of Parliament. Available at: www.legislation.gov.uk/ukpga/2016/1/contents/enacted (Accessed: 29 February 2020).

IAfWM (2017) A new Innovation Alliance for the West Midlands. Innovation Alliance for the West Midlands. Available at: https://birminghamsciencecity.wordpress.com/2017/12/21/a-new-innovation-alliance-for-the-west-midlands/ (Accessed: 29 February 2020).

IAfWM (2018) Innovation Alliance for the West Midlands. Available at: https://birminghamsciencecity.wordpress.com/ (Accessed: 29 February 2020).

Lee, S., Quinn, A. and Rogers, C. (2016) 'Advancing city sustainability via its systems of flows: The urban metabolism of Birmingham and its hinterland', *Sustainability*, 8(3), p. 220. doi:10.3390/su8030220.

LGA (2010) Regional Development Agencies. Available at: www.politics.co.uk/reference/regional-development-agencies (Accessed: 29 February 2020).

Ofgem (2019) Robin Hood Energy Limited: Final order. Ofgem. Available at: www.ofgem.gov.uk/publications-and-updates/robin-hood-energy-limited-final-order (Accessed: 29 February 2020).

Ofgem (2020) Electricity supply market shares by company: Domestic (GB). Ofgem. Available at: www.ofgem.gov.uk/data-portal/electricity-supply market-shares-company-domestic-gb (Accessed: 29 February 2020).

Severn Trent Water (2017) Severn Trent invests £60m in thermal hydrolysis plant to generate more clean energy at Minworth Sewage Treatment Works. News Releases. Available at: www.stwater.co.uk/news/news-releases/severntrentinvestsinthermalhydrolysisplant/ (Accessed: 24 February 2020).

SWM (2015) Our comment: Goodbye to Birmingham 'Energy Savers' and hello 'Energy Leap' … Sustainability West Midlands. Available at: www.sustainabilitywestmidlands.org.uk/news/our-comment-swm-and-members-say-goodbye-to-birmingham-energy-savers-scheme/ (Accessed: 29 February 2020).

Torfing, J. (2005) 'Governance network theory: Towards a second generation', *European Political Science*, 4(3), pp. 305–315. doi:10.1057/palgrave.eps.2210031

Wanner, T. (2015) 'The new "Passive revolution" of the green economy and growth discourse: Maintaining the "sustainable development" of neoliberal capitalism', *New Political Economy*, 20(1), pp. 21–41. doi:10.1080/13563467.2013.866081.

Weaver, M. (2001) 'Neighbourhood renewal: the issue explained', *The Guardian*. Available at: www.theguardian.com/society/2001/mar/13/9 (Accessed: 29 February 2020).

Webb, J., Hawkey, D. and Tingey, M. (2016) 'Governing cities for sustainable energy: The UK case', *Cities*, 54, pp. 28–35. doi:10.1016/j.cities.2015.10.014.

WMCA (2019) The WMCA declares climate emergency and pledges urgent action to reduce emissions. Available at: www.wmca.org.uk/news/the-wmca-declares-climate-emergency-and-pledges-urgent-action-to-reduce-emissions/ (Accessed: 29 February 2020).

4 Case study II: Energy transitions in Frankfurt

Introduction

Emerging interaction patterns among local actors in the cities of Birmingham, Budapest and Frankfurt (visualised and analysed in Chapter 2) revealed that local engagement with the sustainability and low-carbon energy agendas involved multiple actors in each city. The energy transition networks found in the different cities however featured significant variance in terms of structural features.

Frankfurt's network was found to be the largest in size based on the number of member organisations involved in various participative and collaborative decision-making arenas. Despite the apparent wide outreach to external partners, the separation between whole network and the core group indicated a potentially significant network-internal hierarchy between the public sector and the rest of the network. Members of the network's core group – mostly public sector bodies and publicly owned utilities – may have more potential to influence and coordinate network processes compared to Birmingham. In parallel, the market sector is mainly represented collective professional organisations and associations (Chamber of Commerce). Community initiatives, for example energy co-operatives, have also been found to occupy influential positions in Frankfurt's network.

This chapter investigates the local 'context' and its impact on the structural features of Frankfurt's energy transition network (the degree of integration among network actors; network-internal power relationships). The analysis of the context draws on Torfing's (2005) observation that the move away from hierarchies and markets, towards network forms of governance, is facilitated by three key change factors. First, the dispersion of powers, authority and resources in energy systems governance (societal and organisational fragmentation). Second, the locally perceived complexity of energy transitions (complexity in

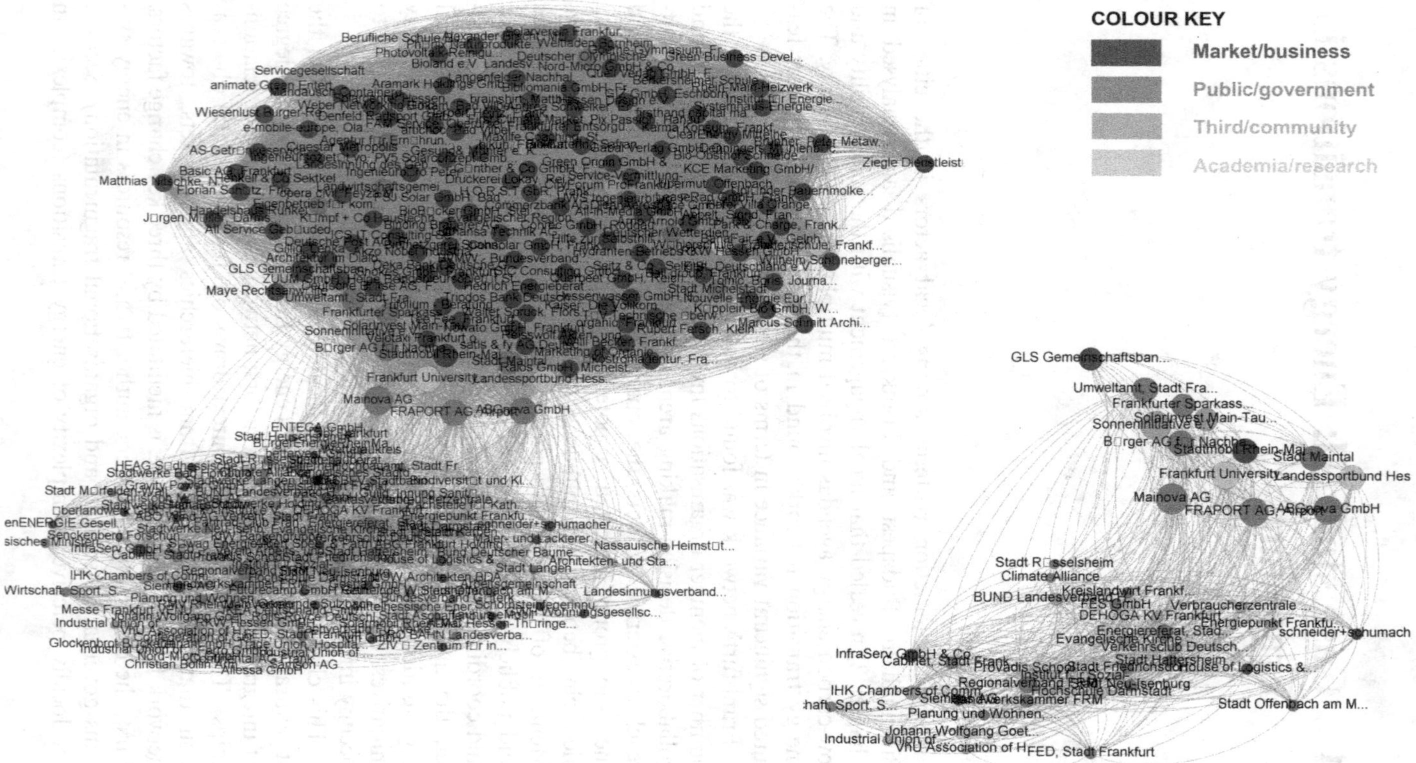

Figure 4.1 The energy transition network in Frankfurt – whole network and core group.

contemporary decision-making). Third, the development of the sustainability agenda over time (new societal dynamics related to a shift from 'government' to 'governance').

The analysis of the complex interplay between contextual change factors and network characteristics can provide insights to guide future attempts aimed at facilitating and coordinating low-carbon urban development through network forms of governance – for example in new TM-backed collaborative transition arenas or through intermediary organisations. Interactions and collaboration within the new arenas are also likely to be influenced by the existing network relationships. Consequently, the analysis of contextual change factors and existing networks can contribute to evaluating the potential of TM to facilitate low-carbon transitions in different cities.

The analysis of the Frankfurt context of low-carbon energy transitions is based on qualitative data obtained from secondary and primary sources. Secondary sources include municipal and partner websites, publications, meeting minutes (publicly available or accessible through FOI requests) and previous studies. Primary data was collected through nine interviews with officers from the Local Authority and their external partners (for a note on the qualitative data collection and a list of interviewees see Appendix I). The interviews focused on questions relating to the role of energy transitions in the local sustainability agenda, the drivers and barriers of transitions, associated governance initiatives and processes, the respondents' involvement in these and the experiences gained, and perceived implications for low-carbon energy transitions in the city.

Frankfurt is situated on the River Main and has a population of just 750 thousand. Its metropolitan area, the Rhein-Main Metropolitan Region, is the second largest in Germany with 5.5 million inhabitants. The city is located near the geographic centre of Europe, and is a major international transportation hub, a centre for trade, culture and education, and the financial capital of mainland Europe (FCC, no date/b). Frankfurt's geographical location and historical Free City ('Freie Stadt') status contributed to its emergence as one of the most influential economic centres, and most wealthy cities, in Europe and internationally.

The city has become an emblematic example of fast economic growth fuelled by the finance industry in the 20th century. Today, its skyline is dominated by over one hundred high-rise buildings, as a result of the boom of the financial industry. Despite the dominance of the financial sector in the city's economy, Frankfurt is also a centre of high-value-added industry and manufacturing (pharmaceuticals, biotechnology) and latterly, digital economy (FCC, 2017). The strong economy, however, leads to high local energy demand, and as a result, high carbon emissions. A key challenge of Frankfurt's energy transition thus relates to finding ways to reduce demand without harming the economy, and meet the reduced needs from local, low-carbon sources.

Energy systems governance in Frankfurt and Germany

Patterns of organisational and societal fragmentation in the governance of energy systems describe the ways which powers and responsibilities are dispersed between different levels of government, and the public, private and civil societal domains, in the energy sector in Germany. This requires investigating how (1) energy policy and regulations are made and by whom, (2) the organisational setup of the ownership and maintenance of energy infrastructures (electricity, natural gas and other decentralised systems), as well as (3) the ways in which energy supply is organised in the German context, and specifically in Frankfurt.

First, in terms of energy policy-making and regulation powers and responsibilities, the organisational landscape is dominated by the federal politico-administrative level in Germany. The formulation of energy policy falls within the remit of the Federal Ministry of Economic Affairs and Energy ('Bundesministerium für Wirtschaft un Energie', BMWi). German energy policy primarily rests on two pieces of legislation: the Energy Industry Act ('Energiewirtschaftsgesetz' 1935/2005) and the Renewable Energy Sources Act ('Erneuerbare-Energien-Gesetz', 2000/2017) (Scholz and Ante, 2018). The Renewable Energy Sources Act provides the legal background to Germany's energy transition ('Energiewende') which states a commitment to a phasing out nuclear energy by 2022, and a 100% shift to renewable energy sources by 2050 (FGG, 2017). Regulatory powers are dispersed between the federal and state levels (Hesse, in the case of Frankfurt) (IEA, 2013). The Federal Network Agency ('Bundesnetzagentur') acts as the main regulatory body. The FNA falls under the BMWi but operates as a separate entity with autonomous decision-making powers. Administrative and accounting duties regarding greenhouse gas emissions fall under the remit of the Federal Environment Agency ('Umweltbundesamt'), part of the Federal Ministry for the Environment, Nature Conservation and Nuclear Safety ('Bundesministerium für Umwelt, Naturschutz and nukleare Sicherheit').

Second, the absence of historical centralisation efforts from the Federal Government resulted in German energy infrastructures (electricity and natural gas) consisting of several main transmission, and regional and local distribution systems (RAP, 2015). Numerous autonomous high-voltage energy and high-pressure gas transmission systems operate within the country, operated by various companies with German or European ownership. Distribution systems are rather decentralised and characterised by a large number of actors: several hundred electricity (appr. 900) and gas (appr. 700) distribution network operators, many of which are relatively small local companies (Gasunie, no date; Appunn and Russell, 2018). In many places, local grids are owned and operated by municipally owned companies, successors to the 'City Works' ('Stadtwerke') model prevalent prior to market liberalisation and deregulation (Scholz and Ante, 2018). The City Works operated as de facto local territorial monopolies before 1998 (RAP, 2015).

Third, energy supply markets are liberalised and operate on the basis of market competition between privately or publicly owned energy producers and suppliers. Competition is overseen by the Federal Network Agency and its regional offices. Centralised power generation in Germany has traditionally been dominated by four large companies (RWE, E.ON, EnBW, Vattenfall), all of which are owned by German or European corporations. Their position has, however, been under treat since the promotion of the energy transition / Energiewende, as they mostly rely on traditional fossil-fuel-based or nuclear generation (Amelang, 2017). Some of the several hundred energy supply companies are also subsidiaries of the 'Big Four' – however, this is not the case in Frankfurt. Frankfurt City Council owns a City Works (Stadtwerke Frankfurt) holding, which includes the local energy company Mainova. The company serves over 80% of local customers (FCC, 2015).

Frankfurt City Council therefore has significant powers when it comes to local energy, mainly through its ownership of Mainova, the municipal energy company. Mainova owns and operates a large number of cogeneration plants (CHP) which run on (mostly) imported fossil fuels. In addition, it generates energy (electricity and biogas) through a waste incineration plant, biomass and biogas plants, and solar panels on municipally owned buildings (including social housing). Solar energy is mainly championed by citizen co-operatives, such as Sonneninitiative and SolarInvest Main-Taunus. Besides the federally coordinated energy transition and Renewable Energy Sources Act, Frankfurt City Council's Energy Agency has been important in bringing new players into the market. The Agency has been providing (free) expert advice and consultancy services to support distributed and renewable production since its establishment in the 1990s.

Complexity and uncertainties of sustainable energy transitions

The origins of the contemporary low-carbon agenda can be traced back to the environmental movements of the 1970s and 1980s in Germany, as well as in Frankfurt. With an economy largely dependent on imported fossil fuels, price volatility and uncertainties in oil supply (created for example by the 1973 oil embargo) hit the country heavily (Planète Énergies, 2015). This period made a significant impact on German energy policy towards prioritising demand reduction and increasing local energy production. A key strategy for demand reduction was the promotion of insulation on buildings to reduce energy demand for heat (Morris and Pehnt, 2012). Local energy generation strategies were based on nuclear production and renewables, especially wind at the time (Planète Énergies, 2015).

Public opinion on nuclear energy, however, soon shifted after the Chernobyl catastrophe in 1986. Strong environmental movements developed which put pressure on the government and the energy sector, demanding a move away from nuclear power and more focus on renewable energy technologies (Papadakis, 2014). This rationale has later been adopted as part of the German energy transition / Energiewende, and the Renewable Energy Sources Act which came into force in 2000. While nuclear power was considered a 'bridging technology' towards a zero-carbon energy system initially, the Fukusima disaster in 2011 reinforced the commitment for the compete phasing out of nuclear power by 2022 (Morris and Pehnt, 2014). Today Germany is one of the main global investors in renewable energy technologies, but also heavily relies on coal (lignite) for energy generation (IEA, 2013). However, continuously increasing energy demand resulting from fast economic growth and a traditionally strong industry / manufacturing sector has meant that, despite the strong support for decarbonisation, emissions remain very high compared to other European countries. In 2017, Germany alone was responsible for 21% of the EU-28's greenhouse gas emissions (Eurostat, 2019).

The energy transition rationale in Frankfurt broadly follows the national pattern. In fact, the city has been one of the areas which saw the fastest economic growth and energy demand increase in the second half of the 20th century. At the same time, it has also been a centre of environmental movements (originating from student protests), and a traditional stronghold of their political representation, The Greens ('Die Grünen', renamed as 'Bündnis 90/Die Grünen' – Alliance 90/The Greens after the reunification of Germany). The often-violent riots of the 1970s between contra-nuclear grass-roots environmental movements and the authorities (Interview 2.03; Papadakis, 2014) led to a relatively strong local societal support for The Greens, ultimately resulting in a rising trajectory both at the state and the federal levels.

For example, one of the emblematic figures of the Frankfurt environmental movements, Joschka Fischer became Vice Chancellor in the Federal Government between 1998 and 2005 in the cabinet of Gerhard Schröder (Social Democratic Party). As Minister for the Environment of the State of Hesse in the 1990s, Fisher championed support for Local Authorities to take up the challenge of 'climate protection' ('Klimaschutz'). Later this initiative was taken over the Federal Government and it still exists today, providing guidance and support for local decarbonisation plans and activities (BMU, 2019).

Enhancing energy security through autarky remains a central driver for the decarbonisation of energy supply in Frankfurt and the surrounding Rhein-Main Region, in line with the federal level narrative. It provides a well-defined focus point for zero-carbon distributed energy beyond the emissions reduction agenda and targets, with a strong societal support. It also contributes to political and public sector leadership for the low-carbon agenda. In Frankfurt, Alliance 90/The Greens have been able to secure constant influence over the City Council's ruling coalitions since the 1990s. Their influence was instrumental in developing cross-party support for the low-carbon agenda, reducing complexity and uncertainty around energy transitions. In line with the broader German context of energy transitions, the local low-carbon energy action is dominated by the public sector, investment banks and citizen initiatives (energy co-operatives; Interview 2.03). Communities' share in low-carbon energy generation nationally accounts to almost half of the installed capacity in contrast with about 5% owned by large energy companies (RAP, 2015).

However, difficulties with reducing energy demand, resulting from continued economic growth (most recently, the boom in digital economy) and the traditionally strong position of energy-intensive sectors (high-value-added manufacturing), considerably weakened the overall effect of low-carbon developments. The city has a commitment to reduce demand by half by 2050 in order to achieve energy independence and zero-carbon supply. In contrast, latest reports from the Energy Agency/Energiereferat show that since the 2008 financial crisis demand continued to rise, and in fact it is higher today than in 1990 (FCC, 2017). Partly resulting from these difficulties, a dominant ideational context developed within the Local Authority that engagement with the market sector does not replace public sector action. Local stakeholders have also been critical of the local level's potential of influencing large industry corporations whose primary market is not confined to the city or national boundaries:

> And all the companies being based here, they do have their own sustainability initiatives and those are very often decided in Switzerland or in the US or in France, where the headquarters are. So there the local debate about what should be done isn't the most influential part. I think this plan [Masterplan 100% Climate Protection] is much more helpful and gives much more guidance to the SMEs, because they can find more innovation fields and so on. And for them, the local market might be more important than the international market.
>
> (Interview 2.05)

Consequently, market sector actors are seen as 'customers' whose energy demand needs to be met – rather than partners in decarbonisation efforts (Interview 2.02; Interview 2.03; Interview 2.05).

The complexity of energy transitions in Germany, and in Frankfurt, thus can be derived from the problematic role of the market sector. Nevertheless, the local public sector's relatively strong position in energy supply in Frankfurt contributes to managing the complexity of energy transitions. Alignment in terms of local, state and federal goals and policy aiming for zero-carbon energy systems by 2050 also provides some certainty over the future of energy supply in Germany. Cross-scale cooperation, specifically with the surrounding Rhein-Main Region, has been identified by Frankfurt's Masterplan 100% Climate Protection (FCC, 2015) as key in delivering low-carbon targets:

> [W]e have to work more closely with the regional side here around Frankfurt. But Frankfurt, of course, it's a magnet for many things, not only for jobs, but also for energy … At the same time, we have very little space here, very limited possibility in terms of working with renewable energies, so it's … clear that to fulfil our targets in a long perspective there has to be a more integrated approach with the region in order to compensate for the limited available space … And I would say that that is clear for everyone.
>
> (Interview 2.07)

The Masterplan sets out the goals of reducing energy demand within the city by half by 2050, and sourcing the remaining energy from renewables sources, including solar, wind and biomass (FCC, 2015). About half of the reduced energy needs of the city will have to come from the region, contributing to a win-win situation by providing jobs and push to the local economies outside Frankfurt.

The historical development of Frankfurt's sustainable energy agenda

The previous sections discuss how the broader political-organisational structure and culture in Germany influences energy transitions in Frankfurt. This section takes a historical perspective and introduces the development of the local sustainability agenda and the associated societal dynamics and changes in the governance of sustainability and low-carbon energy. From the 1990s, in parallel with the rising trajectory of the sustainability agenda in international negotiations, environmental movements and their political representation, Alliance 90/The Greens, have already gained influence in Germany. Figure 4.2 shows an overview of the different governance initiatives with a relevance to the development of the local low-carbon agenda.

In Frankfurt, Greens have continued to be involved in the ruling coalitions of the City Council since 1990 – despite periodic shifts between left- and right-wing leadership. Frankfurt is currently led by the Social-Democrat Lord Mayor, a Christian-Democrat leader of the Council, and a grand coalition of the three strongest parties: the Social Democrats, the Christian Democrats and The Greens. The 1990 coalition was formed between the Social Democratic Party (SPD) and The Greens, with a member of The Greens being appointed as Deputy Mayor for Environmental Affairs (Interview 2.03). Deputy Mayors are elected members of the Council, with their roles assigned through the coalition-forming process. They are members of the cabinet, together with the Lord Mayor, as well as the Leader of the Council (from the leading party, since the Mayor became directly elected rather than appointed).

Upon his appointment, the Deputy Mayor launched a number of initiatives aimed at shifting Frankfurt onto a more sustainable development path. One of these was the Climate Alliance of European Cities with Indigenous Rainforest Peoples (Climate Alliance, no date) which became a major global initiative connecting European cities (as the 'producers' of greenhouse gases) and those most vulnerable to the effects of climate change. The Alliance required signatories to commit to greenhouse gas emissions reduction targets and has been providing space for knowledge- and best-practice-sharing and lobbying. Frankfurt's emissions reduction target was set at 50% by 2010 initially – however, due to a lack of progress, this was revised to 10% reduction every five years until 2030 (Interview 2.02).

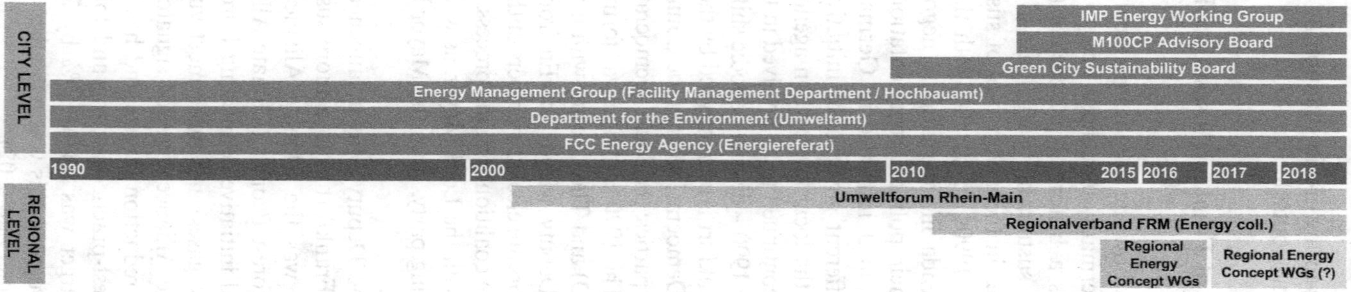

Figure 4.2 Overview of the development of low-carbon energy governance initiatives in Frankfurt.

In order to assist with the delivery of the emissions reduction goals, the Deputy Mayor also established the Frankfurt Energy Agency/Energiereferat within the Department for the Environment/Umweltamt (Interview 2.03). The Energy Agency/Energiereferat has been operating since 1990, led by a Director reporting to the Deputy Mayor for Environmental Affairs. The Agency's work started with developing a robust and reliable methodology for measuring and reporting carbon emissions in the city. Work on assessing the potential of energy saving and energy efficiency measures in emissions reduction has also started. Initial results and recommendations were summarised in Frankfurt's first Energy Strategy/Energiekonzept published in 1997, revised between 2005 and 2007 (FCC, 2019). Key domains for action have been identified as energy saving in buildings, efficiency improvements primarily through heat and power cogeneration (CHP), and renewable energy. The promotion of cogeneration has been particularly successful in Frankfurt, leading to a 'CHP revolution' (Friedel and Neumann, 2002; Transition Cities, 2015). This was facilitated through introducing a local feed-in tariff for CHP (7 cents / kWh), in addition to the Federal Government's renewables subsidies (Interview 2.03). In addition to energy systems and energy efficiency, the Energy Agency/Energiereferat's activities today also include a focus on facilitating behaviour change (e.g. energy saving, food consumption).

Besides the Energy Agency, the Local Authority also has an energy management group within the Estates Management Department/Hochbaumamt, responsible for delivering low-carbon development goals within the Council's own estates. The predecessor of the group was established in 1983, providing assistance to the new Energy Agency. Its main responsibility is to ensure the operations of the City Council provide a 'best practice' example for external stakeholders in the city (Interview 2.02). The group has been advising on passive house construction and refurbishments since the Passive House Resolution was passed by the City Council in 2007. The resolution requires that all new buildings and major refurbishments meet the passive house standard, covering all publicly owned buildings as well as buildings constructed on land acquired from the Local Authority (FCC, 2019). Implementation within the council's own estates is ensured by the municipal housing company, ABG.

Despite some early efforts aiming to engage with external stakeholders, the sustainability agenda of Frankfurt has clearly been dominated by the Local Authority until recently. The first permanent engagement forum, Environment Forum/Umweltforum Rhein-Main was set up in the early 2000s. The forum has been meeting regularly since, and has a large membership of businesses, universities, third sector organisations and public sector bodies (Umweltforum, no date). The Environment Forum/Umweltforum's main purpose is to facilitate networking rather than providing expert advice for policy.

The involvement of external stakeholders in policy consultations started relatively late, in conjunction with Frankfurt's application for the European Green Capital Award 2014/2015. First, a number of citizens' forums and workshops were held throughout the year 2012 around five thematic areas: economy and consumption; planning and building; education; climate and open spaces; and mobility (Umweltamt, no date/b). Outcomes of these engagement processes were subsequently consulted with and evaluated by a Sustainability Board/-Nachhaltigkeitsforum. The forum involved 21 members from public sector including politicians, businesses, professional associations, civil society organisations and academia (Umweltamt, no date/a). Frankfurt's Green City Strategy developed through these collaborative processes earned the city a place among the three finalists of the Award – as well as experience in working with external partners.

Following a similar participative approach, various municipal documents were produced in the subsequent years: Frankfurt City Council published a Masterplan 100% Climate Protection (FCC, 2015), Masterplan Industry (FED, 2016), among other strategic plans to reduce energy demand and carbon emissions (FCC, 2019). Whilst many of the documents are, to some extent, relevant to local energy policy, it was the Climate Protection Masterplan that set out the goal of halving energy demand and satisfying the remaining energy needs from renewable sources by 2050. The Masterplan estimates that this would result in an approximately 95% of carbon emissions reduction (FCC, 2015). The City Council first commissioned an independent research institute (Fraunhofer Institut für Bauphysik, Kassel) to devise various pathways towards achieving this goal. An Advisory Board/ Klimaschutzbeirat was set up to discuss the scenarios proposed by the research institute, led by the Energy Agency/Energiereferat. Members of the Advisory Board were in part identified through a stakeholder mapping exercise (Interview 2.01) and based on the membership of the Green City Sustainability Board (Interview 2.05).

The engagement process for the Masterplan Industry was led by the Frankfurt Economic Development Agency/Wirtschaftförderung, also part of the Local Authority. The Masterplan aimed at developing pathways towards diversifying the economy which became a central concern in Frankfurt following the 2008 financial crisis. The Agency set up eight thematic working groups around eight key themes of industrial site development; digital infrastructure; energy; logistics; industry attractiveness; employment and qualification; administration processes and services; and innovation and value creation of the future (FED, 2016). The Energy Agency was invited to participate in the energy working group. The outcomes of the Masterplan Industry process illustrate well Frankfurt's problem of engaging with the market sector: members of the Masterplan working groups did not agree to any concrete target for low-carbon energy. The Industry Masterplan thus considers such targets on a voluntary basis. More recently, the Energy Agency started a targeted programme aimed at industry parks (e.g. Höchst/Infraserv, Fechenheim-Nord / Seckbach), assigning 'location managers' to help with facilitating energy transitions and reduce emissions.

Based on the recommendations of the Masterplan 100% Climate Protection, work started in 2014 on the regional level (Rhein-Main Region) to draw up plans for meeting half of Frankfurt's future energy demand from renewable sources within the region (FCC, 2015). A collaboration has been set up between the Energy Agency/Enegiereferat and the Regional Authority/Regionalverband in order to coordinate the associated engagement processes. Favourable conditions for collaboration between the city and regions were ensured by the timely appointment of a member of The Greens as Deputy Director of the Regional Authority/Regionalverband. Engagement with external partners at the regional level was carried out through a series of workshops in 2015 focusing on five thematic areas: energy systems, housing, mobility, economy and finance (Regionalverband FRM and FCC, 2016). Outcomes of the workshops were summarised in a Regional Energy Strategy/Regionales Energiekonzept Frankfurt-Rhein-Main. The strategy has identified that much of the required clean energy to satisfy the future demand will have to come from solar power (e.g. '10,000 Solar Roofs Initiative'), and simultaneously expressed the need for implementing distributed energy storage solutions (including energy-to-gas schemes). It also recommends a focus on behaviour shifting, producing an inventory of waste heat availability from industrial processes, and developing a regional energy model (Regionalverband FRM and FCC, 2016).

Outcomes: Decentralised energy in Frankfurt

Local energy generation in Frankfurt is higher than that in Budapest or Birmingham. About 30–35% of the energy demand (electricity and heat) can be met through decentralised production (FCC, 2015), mainly coming from Mainova's cogeneration (CHP) plants, solar power, biogas/biomass, a waste-to-energy plant and block-type CHP installations. In terms of primary energy sources however the situation is somewhat different. Local sources from the Rhein-Main Region cover only about one-fifth of the demand, and only about 5% in Frankfurt (within the city limits; (FCC, 2015)). The reason for this is that main primary energy sources such as natural gas, coal and petroleum products are not available locally and therefore must be imported.

The municipal energy company, Mainova, has the most important distributed generation capacity. It owns, maintains and operates the electricity infrastructure within the city limits, extensive district heating networks, and various heat and power cogeneration plants (CHP). Mainova owns four larger plants which run on natural gas and coal, and four smaller plants – some in partnership with other municipal companies (biomass/biogas, waste incineration). The widespread uptake of block-type CHP units, producing heat and electricity for tower blocks and residential areas which cannot be served by district heat, has been championed by Mainova and the Energy Agency/Energiereferat. Within a few years, several hundred installations have been set up, including one for the offices of the German Federal Bank (Transition Cities, 2015).

About 13% of Mainova's electricity generation is produced from renewables such as large-scale solar, onshore and offshore wind, and hydroelectric power, through investment in power plants outside the region (FCC, 2015). Local solar power and solar thermal installations have been placed on social housing administered by ABG, the municipal housing company. About one-fourth of Frankfurt's population lives in apartments and houses owned by ABG (ABG, no date). In addition, citizens' co-operatives also play an increasingly important role in local (regional) solar power generation – such as Sonneninitiative or Solar-Invest Main-Taunus. The co-operatives rent roof space from building owners, manage the installations and sell the generated electricity to Mainova (Interview 2.04). In 2016, there were about 1,100 installations operating within the city limits (Interview 2.04). In addition, about 1,700 solar thermal installations were in operation in 2015 (FCC, 2015).

The most impactful of the energy saving initiatives has been the Passive House Resolution, and associated assistance for developers from the Energy Agency/Energiereferat and the Council's Energy Management Group/Hochbauamt. In addition, a range of initiatives have been set up to facilitate behaviour change and to provide expert advice on energy saving. Most of these operate as separate entities (e.g.

Umweltlernen, Energiepunkt, 'Frankfurt Saves Electricity'; (FCC, no date/a)) but are overseen and supported by the Energy Agency/Energiereferat. Carried out in collaboration with Caritas (a charity organisation of the German Catholic Church), one initiative explicitly aims to reduce the energy bills of disadvantaged households, and trains long-term unemployed people to become energy consultants (Interview 2.03; Interview 2.07).

As the overview of distributed generation and other associated low-carbon energy projects demonstrates, Mainova and the Energy Agency/Energiereferat are key players in the energy sector in Frankfurt. Whilst Mainova operates as a separate company, the City Council's influence over its operations is ensured not only through ownership but also personal relationships – according to the interviewees. Many of the employees of the city council, specifically the Energy Agency/Energiereferat, and Mainova know one another personally due to the company operating as part of the Local Authority until 1998 (Interview 2.02). The Energy Agency/Energiereferat also supports action on the ground through educational initiatives and providing expert advice to citizens and businesses free of charge. However, links to research and innovation remain somewhat lacking – no major pilot/demonstrator projects have been reported by the key informants.

Networks, context and energy transitions in Frankfurt

German energy systems are characterised by relatively decentralised grid architectures and a large number of actors involved in production, distribution and supply. The energy transition/Energiewende acts a driver for further decentralisation of energy generation, encouraging the spread of small-scale renewable production through feed-in tariffs and other financial support, as well as guaranteed priority connection. Local Authorities, including Frankfurt's, are in a relatively strong position in the energy governance landscape in Germany. This is down to a number of interconnected factors: the absence of centralisation efforts from the federal state; the comparatively late market liberalisation and deregulation in 1998; and the continued dominance of the pre-market competition energy supply culture in which it fell under the remit of Local Authorities.

However, whilst significant capacity and capability for reducing complexity and uncertainty around energy transitions has been developed within the public sector in Germany and in Frankfurt, impact in terms of delivering on policy ambitions has been limited by difficulties to engage with the market sector and gain some control over rising demand from economic growth. For example, Frankfurt's Energy Agency reports that the booming digital economy (about 40 large data centres) have consumed more electricity in 2017 than the total demand of all local households combined (FCC, 2017). This situation contributes to path dependency despite the generally favourable conditions, including political leadership, societal support and the already relatively decentralised grid architectures. In order to tackle this issue, Frankfurt City Council chose to take direct action within its own estates, as well as through its energy company Mainova. In addition, the municipal Energy Agency provides support and resources for citizens and local economic players and other 'non-traditional' players in the energy sector (e.g. financial sector) in order to gain support for decarbonisation goals.

Organisational structures and political commitments (within the Council, as well as at other political-organisational levels) for low-carbon energy have been relatively stable in Frankfurt since the 1990s. The City Council has significant authority over local energy policy, and managed to pass legislation to improve the energy efficiency of buildings (Passive House Resolution) as well as the efficiency of distributed energy generation (CHP feed-in tariff). Investments in solar power is largely driven by federal level policy and subsidy. Whilst the Energy Agency has been working with external partners on a project-basis since its establishment, collaboration and engagement on a more strategic level started relatively recently. However, engagement with the market sector, especially international corporations, remains problematic, leading to difficulties with reducing energy demand and carbon emissions.

The structural analysis of energy transition networks presented in Chapter 2 found that Frankfurt's network was characterised by a large network size and integration among actors from different societal sectors (public, market, third/community). However, the integration of market sector actors was considerably less developed compared to Birmingham's network. These actors remained mainly on the peripheries of the network (e.g. in the Environmental Forum/Umweltforum). Consequently, the core group was characterised by the dominance of the public sector. Together with a greater separation between the network and the core group, this indicated the existence of internal network-hierarchical relationships between local public sector bodies and the rest of the network actors.

The analysis of the local context revealed that lower integration with the market sector results from first, the historically limited role of market players in Frankfurt's energy systems. Second, the shift in energy production ownership from large companies to relatively small-scale generation and new actors entering the market in conjunction with the national energy transition/Energiewende. Third, the strong role of energy-intensive sectors in the local economy, including industry and manufacturing but also digital services. Fourth, overall integration is lower due to smaller and more varied membership of participative decision-making arenas. Power relationships within the network, indicating an influential role for Frankfurt City Council and its Energy Agency/Energiereferat, have been maintained by the Council's authority over the local energy systems, including powers to pass local legislation (Passive House Resolution, local CHP feed-in tariff) and the ownership of the energy company Mainova. Due to the difficulties with engaging with the market sector, a dominant ideational context developed which sees the market sector as customers rather than partners in delivering energy transitions. Consequently, transitions are driven by the public sector, the financial sector (investment banks) and community initiatives and citizen co-operatives. Whilst public sector bodies have been active in the sustainability domain for decades, the move towards more participative decision-making has been slow compared to Birmingham. This lack of historical experimentation with network forms of governance also appears to reduce integration between public and private sector actors.

References

ABG (no date) ABG Frankfurt Holding. Wohnen für alle. Available at: www.abg.de/ (Accessed: 29 February 2020).

Amelang, S. (2017) Battered utilities take on start-ups in innovation race. *Clean Energy Wire*. Available at: www.cleanenergywire.org/dossiers/utilities-and-energy-transition (Accessed: 29 February 2020).

Appunn, K. and Russell, R. (2018) Set-up and challenges of Germany's power grid. *Clean Energy Wire*. Available at: www.cleanenergywire.org/factsheets/set-and-challenges-germanys-power-grid (Accessed: 29 February 2020).

BMU (2019) Die Nationale Klimaschutzinitiative. Nationale Klimaschutzinitiative des Bundesumweltministeriums. Available at: www.klimaschutz.de/die-nationale-klimaschutzinitiative (Accessed: 29 February 2020).

Climate Alliance (no date) Climate Alliance. European municipalities in partnership with indigenous peoples. Available at: www.climatealliance.org/home.html (Accessed: 29 February 2020).

Eurostat (2019) Greenhouse gas emission statistics. Emission inventories. Statistics explained. Available at: https://ec.europa.eu/eurostat/statisticsexplained/ (Accessed: 29 February 2020).

FCC (2015) Masterplan 100% Climate Protection. Frankfurt am Main. Available at: https://frankfurt.de/-/media/frankfurtde/frankfurt-themen/klima-und-energie/pdf/energiereferat-79a/masterplan-climate-protection—summarised-version.ashx.

FCC (2017) CO2-Bilanz Frankfurt-am-Main – Entwicklung und Trends. Available at: https://frankfurt.de/-/media/frankfurtde/frankfurt-themen/klima-und-energie/pdf/energiereferat-79a/meldungen/co2-bilanz-praesentation.ashx.

FCC (2019) Building blocks for climate protection. Available at: https://frankfurt.de/-/media/frankfurtde/frankfurt-themen/klima-und-energie/pdf/energiereferat-79a/bausteine-fuer-den-klimaschutz-2017-und-2018-englisch.ashx (Accessed: 29 February 2020).

FCC (no date a) Energiereferat: Über uns. Stadt Frankfurt am Main. Available at: https://frankfurt.de/service-und-rathaus/verwaltung/aemter-und-institutionen/energiereferat/ueber-uns (Accessed: 29 February 2020).

FCC (no date b) Global market place. Stadt Frankfurt am Main. Available at: https://frankfurt.de/english/about-frankfurt/global-market-place (Accessed: 29 February 2020).

FED (2016) Masterplan Industrie der Stadt Frankfurt am Main. Available at: https://frankfurt-business.net/standort-frankfurt/branchenfokus/masterplan-industrie-der-stadt-frankfurt-am-main/ (Accessed: 29 February 2020).

FGG (2017) Das Erneuerbare-Energien-Gesetz. Federal Government of Germany. Available at: www.erneuerbare-energien.de/EE/Redaktion/DE/Dossier/eeg.html?cm_docId=71110 (Accessed: 29 February 2020).

Friedel, W. and Neumann, W. (2002) Systematic implementation of CHP plants in Frankfurt am Main. Available at: https://docplayer.net/19386645-Systematic-implementation-of-chp-plants-in-frankfurt-am-main.html (Accessed: 29 February 2020).

Gasunie (no date) Infrastructure in Germany. Gasunie Deutschland Transport Services GmbH. Available at: www.gasunie.de/en/infrastructure/infrastructure-in-germany (Accessed: 29 February 2020).

IEA (2013) Energy Policies of IEA countries: Germany 2013. doi:10.1787/9789264190764-en.

Morris, C. and Pehnt, M. (2014) Energy transition. The German Energiewende. Available at: www.energytransition.de (Accessed: 29 February 2020).

Papadakis, E. (2014) *The Green Movement in West Germany (RLE: German Politics)*. Routledge. doi:10.4324/9781315727561.

PlanèteÉnergies (2015) The history of energy in Germany. *Planète Énergies.* Available at: www.planete-energies.com/en/medias/saga-energies/history-energy-germany (Accessed: 29 February 2020).

RAP (2015) Report on the German power system imprint country. Profile report on the German power system. Available at: www.agora-energiewende.de/fileadmin2/Projekte/2014/CP-Deutschland/CP_Germany_update_1015_web.pdf (Accessed: 29 February 2020).

Regionalverband FRM and FCC (2016) Regionales Energiekonzept FrankfurtRheinMain – Das Beteiligungsverfahren … Ergebnisse der Strategiegruppen.

Arbeitsbericht. Available at: www.region-frankfurt.de/media/custom/3255_146_1.PDF?1565682532 (Accessed: 29 February 2020).

Scholz, U. and Ante, J. (2018) Electricity regulation in Germany: Overview. Available at: https://uk.practicallaw.thomsonreuters.com/5-524-0808 (Accessed: 29 February 2020).

Torfing, J. (2005) 'Governance network theory: Towards a second generation', *European Political Science*, 4(3), pp. 305–315. doi:10.1057/palgrave.eps.2210031.

Transition Cities (2015) Maximising Europe's low carbon activities: Moving from individual projects to challenge-led, transition programmes.

Umweltamt (no date a) Frankfurt green city. Nachhaltigkeitsforum. Available at: www.frankfurt-greencity.de/de/vernetzt/netzwerke-kooperationen/nachhaltigkeitsforum/ (Accessed: 29 February 2020).

Umweltamt (no date b) Frankfurt green city. Status and trends. Available at: www.frankfurt-greencity.de/en/status-and-trends/ (Accessed: 29 February 2020).

Umweltforum (no date) Wir über uns. Umweltforum Rhein-Main. Available at: www.umweltforum-rhein-main.de/wir-ueber-uns (Accessed: 29 February 2020).

5 Case study III: Energy transitions in Budapest

Introduction

Emerging interaction patterns among local actors in the cities of Birmingham, Budapest and Frankfurt (visualised and analysed in Chapter 2) revealed that local engagement with the sustainability and low-carbon energy agendas involved multiple actors in each city. The energy transition networks found in the different cities, however, featured significant variance in terms of structural features.

Budapest's network was largely different from the networks of the other two cities (see Figure 5.1). It was considerably smaller and weaker than either Birmingham's or Frankfurt's network. No formalised participative/collaborative decision-making arenas were found; however, a network structure could be reconstructed from semi-formal initiatives and information exchange patterns relating to processes of municipal publishing on a regular basis. Most member organisations thus had a public sector origin (municipal departments, utilities and companies). The network analysis indicated the persistence of public sector internal hierarchies, potentially leading to difficulties with developing stronger horizontal network-type interactions.

This chapter investigates the local 'context' and its impact on the structural features of Frankfurt's energy transition network (the degree of integration among network actors; network-internal power relationships). The analysis of the context draws on Torfing's (2005) observation that the move away from hierarchies and markets, towards network forms of governance, is facilitated by three key change factors. First, the dispersion of powers, authority and resources in energy systems governance (societal and organisational fragmentation). Second, the locally perceived complexity of energy transitions (complexity in contemporary decision-making). Third, the development of the sustainability agenda over time (new societal dynamics related to a shift from 'government' to 'governance').

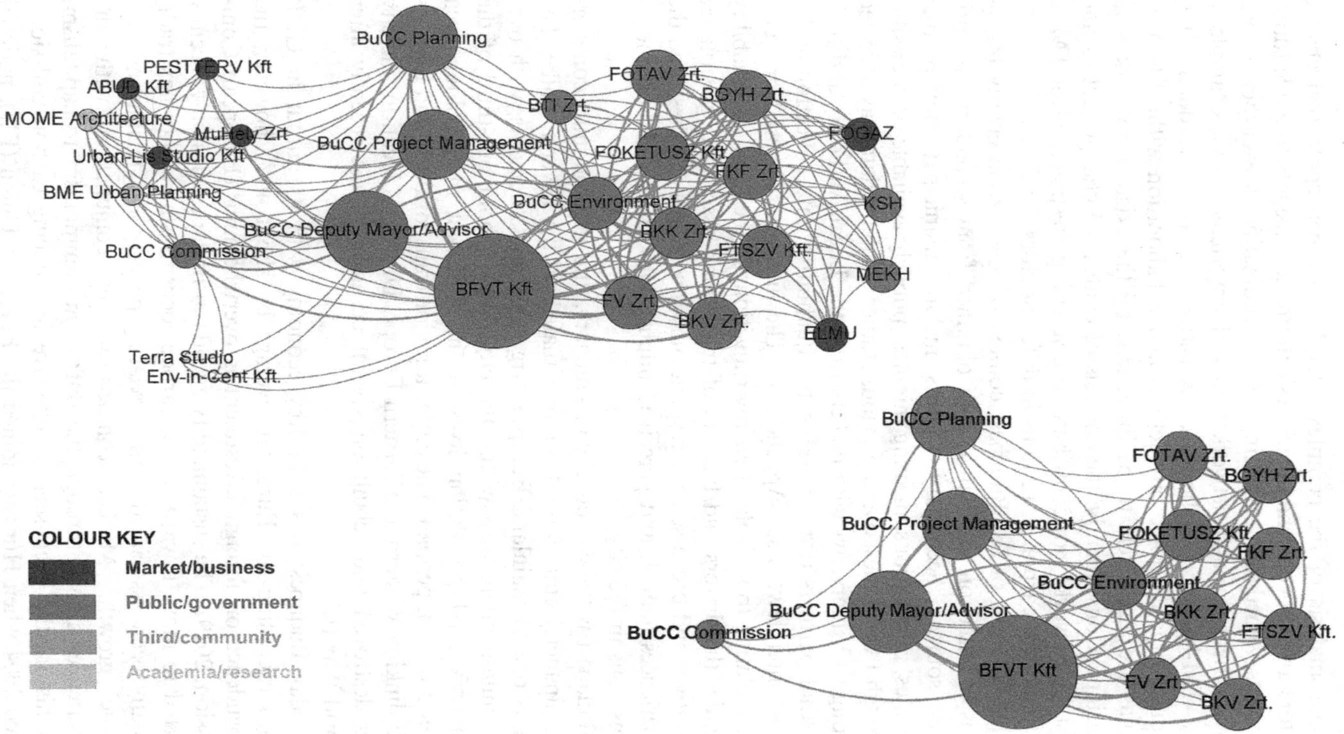

Figure 5.1 The energy transition network in Budapest – whole network and core group.

The analysis of the complex interplay between contextual change factors and network characteristics can provide insights to guide future attempts aimed at facilitating and coordinating low-carbon urban development through network forms of governance – for example in new TM-backed collaborative transition arenas or through intermediary organisations. Interactions and collaboration within the new arenas are also likely to be influenced by the existing network relationships. Consequently, the analysis of contextual change factors and existing networks can contribute to evaluating the potential of TM to facilitate low-carbon transitions in different cities.

The analysis of the Budapest context of low-carbon energy transitions is based on qualitative data obtained from secondary and primary sources. Secondary sources include municipal and partner websites, publications, meeting minutes (publicly available or accessible through FOI requests) and previous studies. Primary data was collected through 12 interviews with officers from the Local Authority and their external partners (for a note on the qualitative data collection and a list of interviewees see Appendix I). The interviews focused on questions relating to the role of energy transitions in the local sustainability agenda, the drivers and barriers of transitions, associated governance initiatives and processes, the respondents' involvement in these and the experiences gained, and perceived implications for low-carbon energy transitions in the city.

Budapest is the capital city of Hungary, and its economic, commercial, financial and cultural centre, situated on the River Danube. The city is one of mainland Europe's largest urban settlements, with over 1.7 million inhabitants within the city limits. About one-third of the country's total population lives in the Budapest metropolitan area (over 3 million people). The city is a major transportation and logistics hub, linking Western and Eastern Europe. Due to its location, the city was destroyed and rebuilt on several occasions, most recently after World War II.

Today Budapest has a healthy economy with its output in GDP terms exceeding the European average by about 40%; however, the dependence on foreign investment is overwhelming (European Commission, 2019). The sustainability agenda in Budapest has only recently started to move towards a low-carbon focus. Sustainability has traditionally been associated with alleviating poverty and facilitating economic growth. Market competition, in conjunction with the experiences gained through the shift from communism to capitalism, has historically been seen as effective in reducing carbon emissions. Since 2004 when Hungary joined the European Union (EU), national and local carbon emissions reduction targets have been following the direction and measures set by the EU.

Energy systems governance in Budapest and Hungary

Patterns of organisational and societal fragmentation in the governance of energy systems describe the ways which powers and responsibilities are dispersed between different levels of government, and the public, private and civil societal domains, in the energy sector in Hungary. This requires investigating how (1) energy policy and regulations are made and by whom, (2) the organisational setup of the ownership and maintenance of energy infrastructures (electricity, natural gas and other decentralised systems), as well as (3) the ways in which energy supply is organised in the Hungarian context, and specifically in Budapest.

First, energy systems (electricity and natural gas) are nationally organised and centrally regulated and operated. Main policy-making responsibilities for energy are assigned to the State Secretary for Energy and Climate Change Policy which falls under the Ministry of Innovation and Technology in the Central Government. Cross-party and cross-departmental coordination, including the Prime Minister's Office, and engagement with external partners, is organised within the frames of the National Commission for Sustainable Development / Nemzeti Fenntartható Fejlődési Tanács. While the Commission has a role in monitoring progress towards sustainability targets, it has no assigned powers or authority. The Hungarian Energy and Public Utility Regulatory Authority / Magyar Energetikai és Közmű-szabályozási Hivatal is the regulatory body of the energy and public utility markets in Hungary, tasked with overseeing market competition in sectors of strategic national importance.

Second, Hungarian electricity and gas systems are centralised as a result of nationalisation and centralisation under the communist regime in the post-war period. The high-voltage electric transmission grid is operated by a single entity, the Hungarian Transmission System Operator / MAVIR, as a natural monopoly. MAVIR is part of the MVM Group which integrates several state-owned energy generators, supply companies and system operators (MVM Group, no date). Downstream distribution networks are operated by six regional distribution system operators which are also the 'universal' (default) suppliers for their respective regions. Most of these are subsidiaries of international energy corporations (E.ON, RWE), but the Central Government has recently shown interest in bringing utility companies back into public ownership via the MVM Group. Gas transmission systems are also operated by a single transmission system operator (Natural Gas Transmission / Földgázszállìtó), part of the MOL Group (MOL Group, no date). The Government is the largest (but minority) shareholder in the MOL Group. The majority of natural gas distribution belongs to five large regional distribution operators (IEA, 2017). The distribution network operators also act as regional universal/default suppliers. Until recently, three of these companies were in foreign ownership. Today, all natural gas regional operators/universal suppliers belong to the Natural Gas Distribution Company / FØGÁZ, owned by the state through its Hungarian Development Bank / Magyar Fejlesztési Bank.

Third, energy systems are liberalised and operate on the basis of market competition; however, as discussed above, the Central Government has a large role in energy supply in Hungary via publicly owned (at least in part) electricity and gas system operators and universal supply companies. Government-mandated end-user price cuts, and the resulting profit losses, have been increasing the role of the public sector in energy supply in recent years (IEA, 2017). The public sector also has a major role in energy generation, e.g. via ownership of the Paks Nuclear Power Plant (IEA, 2017).

Budapest City Council has no involvement in electricity or natural gas supply, and no authority to produce regulation or legislation locally. Regional electricity systems are operated by ELMŰ, a subsidiary of Innogy (multinational company headquartered in Germany). Natural gas is distributed and supplied by FŐGÁZ, a publicly owned company which also acts as the local universal supplier. The Council owns 50% of the shares of the city's public lighting company (BDK) with the other half belonging to ELMŰ. Communication and collaboration with external partners are restricted to sales and emissions data sharing for carbon emissions monitoring purposes (Interview 3.08).

However, through its ownership of utility companies, the City Council has some influence over the energy sector. Municipal utilities include a district heating company (FŐTÁV), and waste (FKF), water (Budapest Waterworks) and sewage (FCSM) management companies. The ownership of the utilities was taken over from the central administration by Budapest City Council during the post-communist reorganisation in the early 1990s. Distributed, low-carbon energy schemes (e.g. waste-to-energy, biogas) are being carried out and managed by the publicly owned utilities. The electricity and heat cogeneration plants feeding the district heating networks are owned and operated by a private company, Budapest Power Plant / Budapesti Erőmű, part of a company called EP Holding, registered in the Czech Republic (Végh, 2016). Encouragement and support for developing small-scale solar power generation capacity is limited both at the national as well as the local levels. In addition to limited opportunities for engaging with the private sector, and artificially low government-mandated unit prices for energy, this situation prevents new actors from entering the energy sector.

Complexity and uncertainties of sustainable energy transitions

During the time of the environmental movements in Germany, Hungary belonged to the Eastern Bloc, a group of satellite states in Central and Easter Europe controlled by the Soviet Union. When the sustainable development agenda became central to international negotiations coordinated by the United Nations in the late 1980s, Hungary had to manage a complete restructuration of its national economic (from the communist to the liberal capitalist model), societal (reinstating the democratic order) and politico-administrative (from centralised to decentralised) frameworks. These experiences have made a significant impact on how the sustainability agenda is interpreted locally, and on the appropriate ways of implementing it. Carbon emissions reduction is generally considered an outcome rather than a goal in itself. Plans and strategies, especially in the energy sector, do not take a radical approach – rather, they aim at decarbonising energy production without changing the grid architectures.

The opportunities offered by the system change, coupled with an enthusiasm towards the neoliberal market-driven development model which proved successful in facilitating fast economic growth in the 'role model' Western countries, contributed to a dominant view that market competition would be sufficient to deliver sustainable development (Kerekes, 2006). Reinstating democracy contributed to enhancing citizens' well-being and improving social sustainability. Market competition was expected to outperform the previous centrally coordinated economy and produce more sustainable economic growth. Shifting the focus of the economy from heavy industries to less energy intensive and polluting sectors provided a solution to combat environmental degradation and reduce pollution. Carbon emission reduction at the time was not a key concern – it was expected that the fundamental changes in the structure of the economy would deliver any commitments made.

In fact, the result was a carbon saving of 43% by 2009 compared to the 1987 baseline, well above the 6% reduction by 2012 required by the Kyoto Protocol (HNG, 2012). In addition, the extensive use of natural gas for heating and hot water production, and the role of nuclear power in electricity supply (accounting for over 50% of electricity generation; (IEA, 2017)), meant that the carbon intensity of the energy sector had already been low compared to other European countries. Initial success in terms of emissions reduction was thus achieved without the need to take a concerted, strategic approach.

The 1 National Climate Change Strategy / Első Nemzeti Éghajlatváltozási Stratégia was only produced in 2008, necessitated by the country's commitment to implement the Kyoto Protocol. It outlined a plan to deliver greenhouse gas emissions reduction commitments until 2025, building on the European goal of 20% reduction by 2020 (NGH, 2008). The Strategy was also informed by the relevant European directives for climate change mitigation due to Hungary joining the European Union in 2004. Whilst acknowledging that previous successes in emissions reduction were due to a decrease in industrial production, and a temporary economic decline during the years of the system change, the 2008 Strategy also stated a priority for facilitating economic growth and reducing poverty. Consequently, it focused on relatively 'cheap' and 'easy' options to comply with emissions reduction commitments rather than developing a vision of radical change (NGH, 2008).

This attitude remained dominant in subsequent strategic plans as well, such as the 2012 National Energy Strategy / Nemzeti Energiastratégia (NGH, 2012), and the 2017 Second National Climate Change Strategy / Második Nemzeti Éghajlatváltozási Stratégia (NGH, 2017). The 2017 Strategy highlighted that by 2014, energy demand increase produced by economic growth balanced out energy savings from efficiency improvements, resulting in shift towards increasing carbon emissions in the subsequent years (NGH, 2017). However, it also states that despite the change in the trajectory of emissions, the country's per capita emissions remained well below the European average, leaving room for manoeuvre.

In parallel with the reduction of market processes' contribution to emissions reduction, the Central Government started to take up a leadership role in facilitating low-carbon development in the energy sector. This is envisaged to be achieved mainly via the modernisation and capacity expansion of the Paks Nuclear Power Plant, large-scale solar power plants and to a lesser extent, geothermal energy (NGH, 2020). The nuclear expansion is, however, highly controversial – the Government argues that it is an effective way of producing carbon-neutral energy, while independent reviews see it as a political decision due to the project's Russian links (Mátyás, 2019). System improvements, including the electricity grid and old district heating networks, are also expected to contribute to reducing demand and emissions in the energy sector.

The complexity of energy transitions is thus managed via avoiding radical change, and taking a 'back seat' approach to carbon emissions reduction driven by a 'not-our-responsibility' attitude. The main driver of low-carbon energy transitions on the national level is decreasing the dependence on energy imports (over 80% of natural gas is imported; (NGH, 2020)). Reducing poverty and facilitating economic growth are also important, supported by government-mandated artificially low energy unit prices (currently the 2nd lowest in Europe) for domestic and small industrial consumers (DN Hungary, 2019).

In this context that is highly politicised and dominated by the Central Government, Budapest City Council decided to abandon the idea of taking up the complex challenge of decarbonising the local energy systems, and instead focuses its efforts on initiatives that can be delivered relatively easily through its utility companies. In line with the national agenda, main components of low-carbon development in the local energy sector include the energy-efficient modernisation of the local district heating networks and public lighting, and energy saving from buildings. In addition, there is a dominant view within the Council that regeneration and urban development also contributes to carbon emissions reduction:

> So, for example, we did not build a fourth underground line to save energy and carbon. But when we were planning 'Metro 4', one of the aspects to consider was how much less CO_2 would be emitted in the city with the completion of this project.
>
> (Interview 3.03)

In an attempt to create better coordination mechanisms between the City Council and the utilities, as well as among the utility companies, the Council has initiated the establishment of an Advisory Board, involved in steering the planned 'Innovation Services Company' (ISCO; (Ernst & Young, 2014)). However, the ISCO is still not operational today, despite being under consideration since 2014. It represents the first effort to integrate stakeholders into decision-making processes in a more network-like setting, albeit only on the implementation/delivery level. Furthermore, participation would initially be restricted to municipal utilities, companies and departments.

The implementation of projects aimed at retrofitting poor quality residential blocks is undertaken by temporary collaborative initiatives set op on a case-by-case basis. Smaller scale projects are generally administered by the relevant district councils with little or no interference from the City Council due to the two-tiered non-hierarchical Local Authority structure in Budapest (Interview 3.06). The City Council provides support for large-scale projects with city-wide impact, and in cases of Council-owned social housing which however only accounts for less that 1% of Budapest's housing stock (Interview 3.03).

The historical development of Budapest's sustainable energy agenda

The previous sections discuss how the broader political-organisational structure and culture in Hungary influences energy transitions in Budapest. This section takes a historical perspective and introduces the development of the local sustainability agenda and the associated societal dynamics and changes in the governance of sustainability and low-carbon energy. Figure 5.2 shows an overview of the different governance initiatives with a relevance to the development of the local low-carbon agenda.

In the 1990s, Budapest's leaders were busy outlining and implementing the changes required or brought about by the system change from communism to capitalism. The associated changes in the central-local government relationships were radical: centralisation and central administration were concepts associated with the communist system. Rather than trying to establish a balanced 'middle' position, city and country leaders opted for extreme decentralisation. In this new era, decentralisation was seen as equivalent to democracy, giving the right of self-government to local-level administrations (Keresztély and Scott, 2012). These changes created a two-tiered, non-hierarchical system of Local Authorities in Budapest, with a central City Council / Budapest Főváros Önkormányzata and 23 district authorities. In the absence of a hierarchical relationship, a division of labour exists between the two different types of local governments. However, these divisions are often unclear and overlapping.

In terms of city planning, the decentralisation also brought procedural changes: instead of Central Government-dictated mid- and long-term strategic plans, yearly budgeting and short-termism became dominant (Tosics, 2006). The lack of city-wide planning and implementation contributed to growing spatial inequalities among city districts developing at faster and slower rates (Kovács and Hegedűs, 2014). Nevertheless, it took over a decade for the City Council to acknowledge the problem and take action. Efforts aimed at balancing

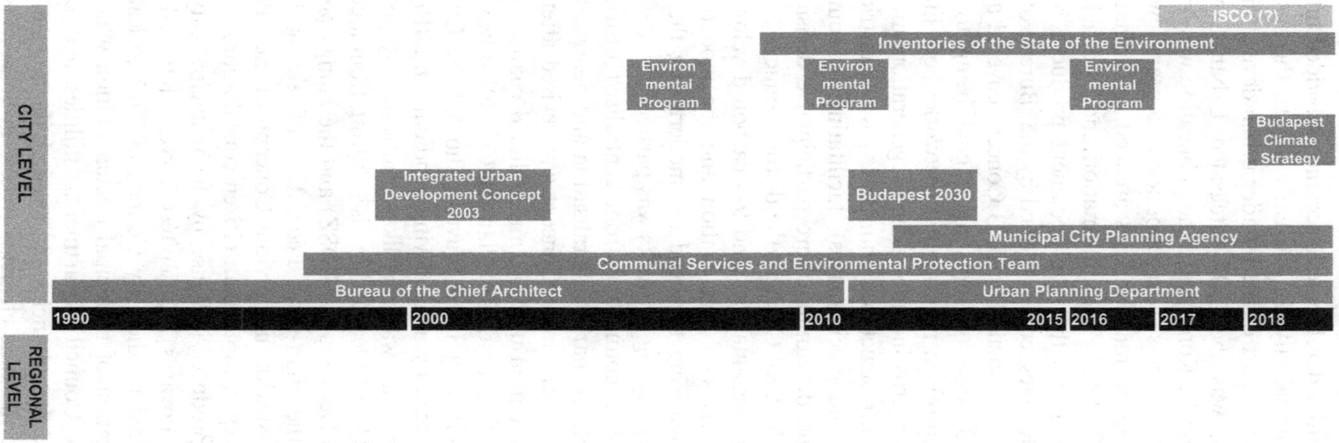

Figure 5.2 Overview of the development of low.

the effects of differential development became embedded into the sustainability agenda of the city. In the early 2000s, the sustainability agenda was discussed as part of the broader urban development plans.

The first of these was the 2003 Integrated Urban Development Concept / Városfejlesztési Koncepció. The publication was preceded by a lengthy consultation process – over a period of 5 years – with citizen groups and experts (academics and researchers), coordinated by the City Council (BuCC, 2003). The information and feedback collected this were incorporated into the final document produced by research institutes and consultancies, commissioned by the 'Bureau of the Chief Architect' of the City Council. The 2003 Concept covered a period of approximately 20–30 years, with its main goal being to reconnect Budapest into the European metropolitan networks both literally and conceptually. Key concerns included urban regeneration and the modernisation of urban infrastructures (transport, sewage; district heating; waste management and green spaces), facilitating economic growth and job creation, and developing more participative decision-making mechanisms across the City Council, the district councils and the citizenry (BuCC, 2003). Sustainability and low-carbon development were thus considered in terms of regeneration and the modernisation of existing assets. Emphasising the need for modernising the inefficient and expensive district heating networks was particularly important due to the popularity of disconnecting from centralised district heating system, made possible by market liberalisation and deregulation.

The 2003 Urban Development Concept was revised after Hungary joined the European Union in 2004, and political changes in the city administration. The Budapest 2030 Plan (BuCC, 2013) was published in 2013, after the left-wing Lord Mayor (who led the City Council since 1994) was replaced by a right-wing candidate in 2010. As such, the new development plan was also likely motivated by political reasons, with the City Council leadership also shifting from a coalition of the Alliance of Free Democrats (SZDSZ) and the Hungarian Socialist Party (MSZP) to the right-wing Federation of Young Democrats (FIDESZ). In contrast with the previous Concept, the new Plan placed more emphasis on expert input than citizen consultations. The majority of the work was produced in-house, by the Municipal City Planning Agency / Budapest Főváros Városépítési Tervező Kft. (BFVT). The company was acquired by the City Council in 2012 to lead the preparation and publication of municipal documentation which requires input from various Council departments, utilities and companies. BFVT commissioned consultancies and research institutes only to lead the work on specific thematic elements (where in-house expertise was lacking). The work of BFVT was overseen by the City Council's Planning Department, replacing the Bureau of the Chief Architect. The draft plan was consulted internally, with the Mayor, his deputies (advisors) and Council departments and municipal companies (restricted to collecting information). Citizen engagement started after the plans were already drawn up – a typical, longstanding practice in decision-making in Budapest, and generally in Hungary (for another example, see Körmendy, 2017).

In terms of content, the Budapest 2030 Plan followed the recommendations and ambitions set out by the previous Urban Development Concept. The novelty lied in establishing extensive information collection processes involving municipal companies and utilities, but also neighbouring Local Authorities, district councils and Central Government departments (BuCC, 2013). The data collected this way were used to describe a baseline situation, and were made publicly accessible. Climate change mitigation and carbon emissions reduction goals and action remained centred around energy-efficient building retrofit and the modernisation of existing infrastructure assets (district heating). The Budapest 2030 Plan also makes a reference to encouraging the uptake of renewables (solar and geothermal) but without any concrete targets or actions. The lack of availability of appropriate funding mechanisms for covering the high costs associated with modernising the old district heating systems was identified as a key obstacle to deliver the goals set out in the Plan. The City Council is highly dependent on the Central Government in terms of finances. Revenues raised from within the city are redistributed across the country, with only about 40% remaining in Budapest – EU funding opportunities and streams are therefore particularly important. This financial dependence on higher levels of government acts a barrier to successful long-term strategic planning (Interview 3.05).

The data gathering processes developed in conjunction with the Budapest 2030 Plan have since become 'institutionalised', with annual progress reports produced by BFVT, overseen by the Communal Services and Environmental Protection Team / Kommunális Közszolgáltatási és Környezetügyi Osztály within the Asset Management Department / Városigazgatóság Főosztály of the City Council. The Team is also responsible for managing service provision contracts with the municipal utilities and companies (Interview 3.02). The series of monitoring documents is entitled Inventory of the State of the Environment of Budapest / Budapest Környezeti Állapotértékelése. The documentation has been published every year since 2011 (BuCC, 2017b). The Inventories also support the bi-annual reporting process of the Covenant of Mayors for Climate & Energy initiative which

Budapest joined in 2008 (CoM, 2012). As a signatory of the Covenant of Mayors, the city committed itself to reduce carbon emissions by 21% by 2020, and 40% by 2030. As part of the signatories' commitments, the City Council produced a Sustainable Energy Action Plan in 2008, which became part of the subsequently produced Budapest Environmental Program 2011–2016 (BuCC, 2011).

The data collection and progress monitoring processes also provide input for the Budapest Environmental Programs / Környezeti Program which are produced every five years. These short-term plans contain more detailed and action-oriented information on the city's environmental agenda. The Environmental Programs link long- and mid-term plans to implementation through identifying core areas for action, potential interventions and funding options to finance delivery (BuCC, 2011, 2017a). As such, they do not introduce new directions or long-term plans outside the scope of the other documentation published by the City Council. Environmental Programs were prepared since 2002. The 2007 Program was produced via a similar engagement process as the 2003 Urban Development Concept – even a formal body has been established which allowed for participation from citizens named Urban Forum for Environmental Consultation / Fővárosi Környezeti Konzultációs Fórum (Greenfo, 2007). The 2007 Program was also revised by the incoming FIDESZ leadership, resulting in a new Program published in 2011. This new Budapest Environmental Program 2011–2016 was produced by an external consultancy (Env-in-Cent Kft.), overseen by the City Council's Urban Development, Transport and Environmental Affairs Commission / Városfejlesztési Közlekedési és Környezetvédelmi Bizottság, set up in 2010 (BuCC, 2011). Due to the Environmental Programs' aim of linking interventions to funding, the Commission have also had a significant role in funding allocation through the Environmental Protection Fund / Környezetvédelmi Alap (Interview 3.05). The Commission consists of councillors from the City and District Councils, representing various political parties and is still currently active. It has no members from the executive branch of the Council or external partners. The current Environmental Program 2017–2021 has been produced by BFVT (with some input from the consultancy producing the previous version, Env-in-Cent Kft) (BuCC, 2017a). This new Program specifically highlights that climate change and environmental protection goals must be integrated into existing activities of the Council rather than discussed in isolation (e.g. urban development, operations, energy and transport management).

Most recently, the City Council published a Climate Change Strategy / Klímastratégia (BuCC, 2018). The document was prepared by

BFVT, overseen by the Investment and Project Management Department. The Strategy includes a new commitment from the City Council, which joined the Under2 Coalition (Under2, 2020) and accepted the target associated with this membership, setting a long-term emissions reduction goal of 80% by 2050. Albeit this commitment seems ambitious, in reality it fits into the national level commitment of achieving carbon neutrality by 2050. The 2018 Strategy identified three core themes for emissions reduction: improving the energy efficiency of buildings, industrial processes and infrastructures, reducing emissions from transport, and improving green area coverage (BuCC, 2018, p. 86). Facilitating behaviour change both within the City Council and among citizens has also been mentioned as a key priority; however, concrete action on how this would be delivered remained lacking.

In conjunction with the Climate Change Strategy, the City Council has been planning to establish a formal collaborative initiative among its departments, utilities and companies to drive low-carbon innovation and enable the development of cross-cutting projects and initiatives. The concept of the so-called Innovation Services Company (ISCO) was developed on the basis of the popular Energy Services Company (ESCO) model (Ernst & Young, 2014). The key responsibilities of the ISCO have been formulated around core themes such as integrating innovation action in the fragmented public service system of Budapest, producing and overseeing a more strategic approach to synergistic service improvement, and identifying funding for delivery (Ernst & Young, 2014). Seed funding was to be made available through setting up a so-called Urban Development Fund / Városfejlesztési Alap, to be complemented via funding from other sources (Interview 3.05). Innovative projects were to be prioritised, as well as collaboration between municipal utilities and companies and the public sector (e.g. startups, small- and medium-sized companies; Interview 3.05). In structural terms the ISCO was envisaged to be set up as a separate entity. Its operations would be overseen by the City Council and an Advisory Board involving senior officers, councillors, and representatives of Council departments, municipal companies and utilities. Over time, the Advisory Board would be extended to include external partners such as research institutes and consultancies (in a second phase) (Interview 3.09).

Despite being on the policy agenda since 2014–2015, the ISCO and the relevant organisational structures and funding mechanisms are still not operational today. In the meantime, the political leadership of the City Council shifted as the previous right-wing Mayor was replaced by the left-wing coalition candidate. Consequently, it is uncertain whether either of these plans will be implemented in the future. Nevertheless, the interactions among the organisations and municipal bodies, which are planned to be formalised and strengthened through the ISCO Advisory Board, exist already.

Outcomes: Decentralised energy in Budapest

According to the latest figures, about 20% of Budapest energy demand is generated locally, which is mainly heat (33% of the total heat demand) and some electricity (about 15% of the total demand) (BuCC, 2017b). This is mainly covered by heat and power cogeneration at the plants owned and operated by Budapest Power Plant / Budapesti Erőmű (private company) which run on natural gas. The municipal district heating company, FŐTÁV is only responsible for the operation of the district heat networks. In addition, a waste-to-energy scheme, operated by the municipal waste management company, FKF, produces energy locally. The waste incineration plant underwent modernisation in 2005 and as a result today operates with heat and power cogeneration technology, using about two-thirds of the non-recyclable waste of the city (BuCC, 2017a).

Plans have been drawn up for a second waste-to-energy plant which would be developed as a joint effort by FŐTÁV, FKF and the municipal wastewater management company, FCSM. However, they were put aside partly due to public complaint and partly to shift in Central Government focus towards modernising existing assets, such as retrofitting lignite power plants to run on solid waste and sewage sludge (Somogyi, 2019).

Budapest has a large potential to use readily available geothermal heat, for example the waste heat from ten operating thermal baths within the city limits. One project has recently been completed, providing carbon-neutral heating for the majority of the buildings of Budapest Zoo. This was carried out in collaboration among the municipal companies: Budapest Zoo, BGYH (thermal baths owner and operator company) and FŐTÁV. There are 118 natural and constructed hot water springs in Budapest, delivering approximately 70 million litres of hot water daily with a temperature between 21°C and 78°C – the majority of which is currently not utilised (BGYH, no date).

The utilisation of solar power is rather underdeveloped in Budapest, due to a lack of leadership from the Council. Energy price caps and low feed-in tariffs act as a barrier for the market sector to drive roll-out. Less than 1% of the local electricity demand is currently met through solar power, most of which has been installed as part of the subsidised energy-efficient modernisation of poorly insulated prefabricated Soviet-style housing blocks built in the 1960s and 1970s (BuCC, 2017b; Interview 3.06). Various programmes and preferential loans are available which aim at providing support to building owners, for example the 'National Energy Conservation Programme', the 'Green Investment Scheme' and the 'Panel-Block Apartment Programme'. These projects are administered by the Central Government, with assistance from the City Council and the district councils.

All documents produced by Budapest City Council highlight the importance of facilitating behaviour change, providing information and raising awareness around issues such as climate change and its local effects, energy use and waste management. However, concrete initiatives and engagement, as well as leadership from the City Council, are currently lacking.

Networks, context and energy transitions in Budapest

In Hungary, and specifically in Budapest, energy systems are dominated by the Central Government and publicly owned companies, and to some extent international corporations (e.g. ELMŰ/Innogy, BERT / EP Holding). Energy policy and decarbonisation plans are also driven by the Central Government, leaving little room for manoeuvre for Budapest City Council. The Council's options to influence local energy transitions are thus limited, and mainly rest with municipal district heating company (FŐTÁV). There are no major commitments from the national level for further decentralisation, preventing new entrants from joining the market.

So far there has been no need to develop new capacity for reducing complexity and uncertainty around energy transitions either at the national level or locally, in Budapest. Impact is being delivered mainly through efficiency gains and the modernisation of existing infrastructures, rather than via restructuring infrastructure systems. This situation contributes to path dependency due to a lack of need and commitment for initiating considerable changes in the current energy systems. Carbon emission reduction targets, following European and international commitments rather than local ambitions, so far appear to be within reach even in the absence of large-scale transitions. In this context, Budapest City Council have avoided taking up a leadership role, beyond meeting the requirements set by the national level. Initiatives involve only a low number of actors, over whom the Council has oversight due to ownership relationships – contributing to a lack of opportunities for new local actors to enter the energy sector.

Climate change mitigation targets are set based on external commitments in Budapest, following the national level example, rather than citizen engagement and local public support for low-carbon transitions. Interaction with external partners and citizens is extremely limited, especially since 2010 – it is yet to be seen whether the new left-wing leadership introduces any change. The Council's main levers over carbon emissions reduction from the energy sector are through its ownership of utility companies. However, collaboration and cooperation across the Council, the utilities and other municipal companies is still lacking even decades after the ownership and oversight of the organisations has been taken over by the Local Authority from the central administration. The companies operate as separate entities, making it difficult to set up and implement cross-cutting projects. A type of coordination role is overtaken by BFVT, the Council's Urban Planning Agency, operating as a separate entity with no formal decision-making authority. The City Council has so far not undertaken any steering role in the network. Weak network relationships, and the lack coordination is likely to be one of the reasons for slow progress with implementation and delivery despite the regularly produced municipal plans, strategies and progress reports.

The above reasons result in a lack of leadership from the City Council, and limited engagement with the market sector and citizens. The structural network analysis highlighted this lack of integration of external partners via participative decision-making initiatives. Consequences include the small and exclusive network consisting of public sector bodies and municipal companies only, and relatively low levels of integration among network actors (density). The lack of diversity within the network contributes to maintaining power relationships within the network, based on public sector internal hierarchies. The context analysis revealed that cross-sectoral integration is restricted not only due to a lack of legal framework for integrating external partners into public sector decision-making but crucially to the lack of pressure and perceived need to reorganise the existing energy systems. In addition, integration across municipal departments and companies in Budapest is hindered by a historical independence as utilities were run by the central administration prior to the 1990s. Integration among policy decision-making and investment on the one hand, and delivery/progress monitoring on the other also appear to be underdeveloped, and interactions remain based rather on traditional hierarchical relationships.

References

BGYH (no date) Budapesti Gyógyfürdők – Termálfürdők Budapesten – Budapesti Fürdők. Available at: www.budapest.com/egeszseg_wellness/budapesti_gyogyfurdok.hu.html (Accessed: 29 February 2020).

BuCC (2003) Budapest Városfejlesztési Koncepció – összefoglaló. Available at: www.urbanisztika.bme.hu/segedlet/bp_fuzet/Urban-Development-Concept-of-Budapest.pdf (Accessed: 29 February 2020).

BuCC (2011) Budapest Environmental Program 2011–2016. Available at: https://budapest.hu/Lapok/Hivatal/Kornyezetvedelem.aspx (Accessed: 29 February 2020).

BuCC (2013) Budapest 2030 Plan: Long-term urban development concept. Available at: https://budapest.hu/Documents/Városépítési Főosztály/Budapest2030_ENG_summary.pdf (Accessed: 29 February 2020).

BuCC (2017a) Budapest Environmental Program 2017–2021. Available at: https://budapest.hu/Lapok/Hivatal/Kornyezetvedelem.aspx (Accessed: 29 February 2020).

BuCC (2017b) Budapest Környezeti Állapotértékelése. Available at: https://budapest.hu/Lapok/Hivatal/Kornyezetvedelem.aspx (Accessed: 29 February 2020).

BuCC (2018) Budapest Klímastratégiája. Available at: https://budapest.hu/Documents/klimastrategia/Bp_Klimastrategiája_vegleges_KGYelfogadott.pdf (Accessed: 29 February 2020).

CoM (2012) Covenant of Mayors – Signatories: Budapest. Available at: www.covenantofmayors.eu/about/covenant-community/signatories/overview.html?scity_id=11938 (Accessed: 29 February 2020).

DN Hungary (2019) Hungary household energy prices among lowest in Europe. Available at: https://dailynewshungary.com/hungary-household-energy-prices-among-lowest-in-europe/ (Accessed: 29 February 2020).

Ernst &Young (2014) Study on the implementation of the ISCO scheme. Available on request from Budapest City Council.

European Commission (2019) Budapest – Internal market, industry, entrepreneurship and SMEs. Regional Innovation Monitor Plus. Available at: https://ec.europa.eu/growth/tools-databases/regional-innovation-monitor/base-profile/budapest (Accessed: 29 February 2020).

Greenfo (2007) Budapest Környezeti Programja 2007. Greenfo. Available at: https://greenfo.hu/hir/budapest-kornyezeti-programja-2007_1190871084/ (Accessed: 29 February 2020).

IEA (2017) Energy Policies of IEA Countries: Hungary 2017 Review. Available at: https://webstore.iea.org/energy-policies-of-iea-countries-hungary-2017-review (Accessed: 29 February 2020).

Kerekes, S. (2006) 'Fenntarthatóság Közgazdasági Értelmezése', in M. Bulla and P. Tamás (eds), *Fenntartható Fejlõdés Magyarországon, Stratégiai kutatások – Magyarország 2015*, pp. 196–211. Új Mandátum Könyvkiadó.

Keresztély, K. and Scott, J. W. (2012) 'Urban regeneration in the post-socialist context: Budapest and the search for a social dimension', *European Planning Studies*, 20(7), pp. 1111–1134. doi:10.1080/09654313.2012.674346.

Körmendy, I. (2017) Utóirat az olimpiáról. Available at: https://epiteszforum.hu/utoirat-az-olimpiarol (Accessed: 29 February 2020).

Kovács, Z. and Hegedűs, G. (2014) 'Gated communities as new forms of segregation in post-socialist Budapest', *Cities*, 36, pp. 200–209. doi:10.1016/j.cities.2013.08.004.

Mátyás, E. (2019) Nuclear trap: Past and future of Hungary's nuclear policy in the light of the Russian influence. Available at: https://energiaklub.hu/en/study/nuclear-trap-past-and-future-of-hungarys-nuclear-policy-in-the-light-of-the-russian-influence-4702 (Accessed: 1 March 2020).

MOL Group (no date) Natural gas transmission. MOL Group. Available at: https://molgroup.info/en/our-business/natural-gas-transmission/natural-gas-transmission-1 (Accessed: 29 February 2020).

MVM Group (no date) Company history. MVM Group. Available at: http://mvm.hu/organization-of-mvm/company-history/?lang=en (Accessed: 29 February 2020).

NGH (2008) Első Nemzeti Éghajlatváltozási Stratégia 2008–2025. Available at: www.terport.hu/teruletfejlesztes/orszagos-szint/fejlesztesi-dokumentumok/nemzeti-eghajlatvaltozasi-strategia-2008-2 (Accessed: 29 February 2020).

NGH (2012) Nemzeti Energiastratégia 2030. Available at: https://2010-2014.kormany.hu/download/4/f8/70000/Nemzeti%20Energiastrat%C3%A9gia%202030%20teljes%20v%C3%A1ltozat.pdf (Accessed: 29 February 2020).

NGH (2017) Második Nemzeti Éghajlatváltozási Stratégia. Available at: www.kormany.hu/download/f/6a/f0000/NÉS_2_stratégia_2017_02_27.pdf.

NGH (2020) Klíma – és energiastratégiai dokumentumok. Available at: www.kormany.hu/hu/dok?source=11&type=402#!DocumentBrowse (Accessed: 29 February 2020).

Somogyi, O. (2019) Egyelőre nem épül másik hulladékégető erőmű. *Magyar Nemzet*, 25 June. Available at: https://magyarnemzet.hu/gazdasag/egyelore-nem-epul-masik-hulladekegeto-eromu-7053587/ (Accessed: 29 February 2020).

Torfing, J. (2005) 'Governance network theory: Towards a second generation', *European Political Science*, 4(3), pp. 305–315. doi:10.1057/palgrave.eps.2210031.

Tosics, I. (2006) 'Nagyvárosok Fenntartható Fejlõdése és Budapest Jövõképe', in M. Bulla and P. Tamás (eds), *Fenntartható Fejlõdés Magyarországon, Stratégiai kutatások – Magyarország 2015*, pp. 420–443. Új Mandátum Könyvkiadó.

Under2 (2020) Our members. Under2 Coalition. Available at: www.under2coalition.org/members (Accessed: 29 February 2020).

Végh, Z. (2016) NRGreport – InterjúErőműveket vásárolhat a Budapest Erőmű Zrt. Available at: http://nrgreport.com/cikk/2016/06/06/eromuveket-vasarolhat-a-budapest-eromu-zrt (Accessed: 29 February 2020).

6 Contextualising energy transition networks

Introduction

'Contextualisation' is a central issue in cross-national comparative research in politics and public administration, which so far has been given little consideration in the literature on governance networks and sustainability transitions. The term refers to the assumed relations, and their importance, between the phenomenon under investigation and the environment – 'context' – in which it is embedded. Research looking for patterns and generalisable conclusions treats context as being irrelevant to the subject of the enquiry. On the other end of the spectrum, culturalist approaches focusing on differences and uniqueness claim that social phenomena are inseparable from the context in which they take place (see for example Hantrais, 2013). As such, these approaches reject the possibility of meaningful comparison across different settings (for example national or sub-national).

In between the two extremes, a variety of intermediate positions have been developed which acknowledge the potential complexity of the relations between the context and the studied phenomenon, but also contend that in real-world settings options in terms of variance are likely to be limited (Faure, 1994). Consequently, developing an understanding of which contextual factors account for the observed variance is central to these research approaches. Context is considered here as an enabling tool for comparison, and part of the explanation, with the studied phenomenon being partially dependent on the context (Hantrais, 1999).

The study of the energy transition networks in Birmingham, Budapest and Frankfurt found governance networks whose remit included local sustainable energy systems in all three cities (see Chapter 2). These existing networks, however, exhibit significant variance in terms of network size, nature (formalised/semi-formalised), the degree of integration (cohesion) and power relationships among network actors across the different cities. In order to unravel potential causes that can explain similarities and differences across the networks, Chapters 3, 4 and 5 discussed the local 'environments' for energy transitions focusing on specific contextual factors that had been linked to the emergence of the network concept in public policy and implementation (Torfing, 2005).

This research thus takes an intermediate position regarding the assumed relations between the studied phenomenon (governance networks) and the context (the socio-technical environment in which they are embedded). The network analysis demonstrated that concepts of 'governance' and 'networks' are potentially generic enough to enable meaningful comparison across the three cases (Pierre, 2005). However, the observed similarities and differences also point to important considerations relating to the context-dependence of 'changes in urban bureaucracy such as the move towards a blurring of public-private boundaries, the rise of an increasing number of governance networks, and a greater inclusion of actors other than the local state in the pursuit of community goals' (Kjaer, 2009, p. 137).

The comparison of energy transition networks in Birmingham, Frankfurt and Budapest investigates how specific contextual characteristics influence the emerging network structures and processes, and their role in governing sustainable energy transitions, in each city. The findings are relevant to theories and frameworks that promote network forms of governance to facilitate urban sustainability transitions – including the Transition Management and intermediation models. They highlight the need for developing a more nuanced understanding of governance structures and processes in networks operating in different urban settings, based on influential local contextual features. Such an understanding is key to evaluating the potential of network forms of governance to facilitate and coordinate low-carbon transitions.

How does context matter? Comparing energy transition networks in Birmingham, Frankfurt and Budapest

Chapter 1 ('Network governance in sustainability transitions') argued for the importance of considering context-dependence in the case of networked forms of governance in order to assess their potential contribution to facilitating urban sustainability transitions. The argument rested on results of previous research in comparative politics and public administration investigating how different national and local administrations in Europe (and internationally) made sense of, and reacted to, historical developments in the 20th century (Pollitt and Bouckaert, 2011). These include for example the emergence of waves of reforms in the public sector (relating to the functioning of the state and its role vis-à-vis other societal domains), the changing European and international context (globalisation, and 'Europeanisation' in the European Union), and the growing number of complex problems which are difficult to tackle within the contemporary (inherited) political, organisational and economic frameworks. The diversity of approaches to encounter these issues nationally and locally have been shown to result in variance in terms of both local governance network structures and processes, as well as regarding the role of networks within the broader governance arrangements of urban sustainability transitions (cf. Skelcher, Sullivan and Jeffares, 2013).

The comparative analysis of network structures uncovered apparent differences among the energy transition networks in Birmingham, Frankfurt and Budapest. These differences related to integration (network size, density, types of actors involved) and power relationships (sectoral dominance, actor centralities, openness) within the networks. More similarities could be identified between the networks of Birmingham and Frankfurt, whilst Budapest's network stood out both in terms of relatively low levels of integration and significant power differences among network actors. The analyses of the local contextual factors in each city were structured around factors relevant to the development and functioning of governance networks. These included local patterns of societal and organisational fragmentation, interpretations of complexities and uncertainties pertinent to energy transitions, and the changing societal dynamics in relation to the development of local sustainable energy agendas (Torfing, 2005). In the following, the similarities and differences among the energy transition networks in terms of integration and power relationships will be connected to these factors and their impact.

The impact of the 'context' on network integration

'Integration' here refers to issues of network cohesion and connectedness (density) and outreach to actors representing different societal domains (size, types of actors involved). A higher degree of integration within the network indicates a move towards network governance as the potentially more dominant mode of coordination. According to the context analyses, the following characteristics have been found to be influencing network integration:

1 The distribution of resources (authority) across different political-organisational scales and different societal domains (public and market sector, community) in terms of societal and organisational fragmentation.
2 The urgency and scale of change required to deliver on carbon emissions reduction goals compared to currently existing socio-technical systems of energy (including physical infrastructure and institutional-organisational structures) relating to complexity and uncertainty.
3 The accumulated experience with participative/networked forms of governance relating to changing societal dynamics and the shift from 'government' to 'governance'.

First, characteristics of societal and organisational fragmentation contribute to varying levels of integration in the energy transition networks based on the patterns of resource distribution vertically, among different political-organisational scales – and horizontally, among actors from the public, market and third sectors (societal and organisational fragmentation). In terms of vertical resource distribution, all three cities are, to some extent, dependent on higher levels of government as policy-making and regulatory powers in the energy sector are concentrated at the national scale. Differences, however, materialise based on the Local Authorities' capacity and capability to manage local energy systems.

In Birmingham, such capacity and capability are relatively weak as the Council has no role in operating or managing local energy infrastructures. Frankfurt City Council's stronger position is based on its municipal energy company Mainova, and expert knowledge accumulated within different Council departments, including the Energy Agency. In Budapest, capacity and capabilities are associated with the municipal utilities (mainly district heating and waste management) and, to a lesser extent, internally with the Council's Environmental Protection Team. The ways in which market liberalisation has been implemented historically in the respective national contexts (and locally) has implications for horizontal resource distribution patterns. The UK has been one of the pioneering countries in opening up the energy sector to market competition and opted for widespread privatisation of previously publicly owned assets and infrastructures. Consequently, private sector companies play a dominant role in energy production, distribution and supply. This results in a comparatively weak position of Local Authorities in the energy systems governance landscape, which in the case of Birmingham leads to problems with coordinating network processes and outcomes. Germany introduced market competition in the energy sector around the millennium, but the legislation allowed for the continued operation of municipal energy companies as publicly owned but separate entities. This allows Frankfurt City Council to retained oversight and influence over energy production, distribution and supply via Mainova, and other municipal companies (waste management, social housing). Hungary implemented market competition in a similar vein to Germany. Publicly owned companies ('universal suppliers') are therefore in a strong position – however Budapest City Council does not own or have jurisdiction over the regional universal supplier. Local energy generation is also outside the reach of the Council, and therefore influence is limited to municipal utilities (district heating, waste management).

Second, the necessity of governing urban sustainability transitions via network forms of governance is derived from their 'default' characterisation as 'complex' problems – both in Transition Management and the intermediation model. However, the locally perceived nature and degree of complexity and uncertainty of energy transitions varied across the three city cases. The 'urgency' and scale and nature of change required to deliver on low-carbon targets have been found to be a key in increasing the need for networked forms of governance and integration. In Birmingham, an ambitious emissions reduction goal (urgency) at the time exceeding the national target, coupled with the slow progress in developing decentralised energy systems and local expertise (scale/nature), corresponded to higher levels of integration compared to the other two cities. The combined effect of these factors, and the lack of options available to the City Council to directly influence the pace and trajectory of change, resulted in a dominant view that engagement with external partners was necessary to manage the complexities and uncertainties of low-carbon energy transitions and achieve the emissions reduction goals. In Frankfurt, emissions reduction targets were found to be ambitious, but in line with the national goals. However, an additional requirement was set to achieve these goals through using only locally generated energy. This led to initiating the integration of actors from the local and regional scales, and a distinct focus on new entrants to the energy sector (co-operatives, investment banks), in order to develop capacity for managing the complexities and uncertainties associated with the envisaged transition path towards low-carbon energy systems. In Budapest, resource efficiency gains achieved through the introduction of market competition, the energy efficient modernisation of existing infrastructures and building stocks, complemented with the national-level efforts to decarbonise electricity production, have so far been sufficient to deliver the carbon emissions reduction goals set by the City Council. Consequently, from the perspective of Budapest City Council, complexities and uncertainties associated with low-carbon energy transitions were considered relatively low and engagement with, and outreach to, new actors has so far been not required.

Third, more experience with participative and collaborative forms of governance along the historical development of the broader sustainability (and sustainable energy) agenda have been found to correspond to higher levels of network integration. Birmingham has had the longest history of participative, networked forms of governance in the sustainability domain. Involvement in various initiatives historically has allowed for more integration among network actors over time – also referred to as 'institutionalisation' in the network literature (Torfing and Sørensen, 2014). In Frankfurt, formal participative initiatives (beyond citizen engagement) are fairly new phenomena compared to Birmingham. Instead, the sustainability agenda has historically been led by the City Council, providing support and best-practice example for external stakeholders. Interactions among network actors in Budapest have so far not led to formal participative and collaborative decision-making initiatives. At the time when the sustainability agenda first gained traction, Hungary was undergoing a system change from the communist to the capitalist model. Consequently, early collaborative efforts (e.g. citizen forums relating to the Local Agenda 21 movement) were lacking in Budapest. Similar legal frameworks that could support setting up such multi-actor arenas are also lacking – in the absence of such legal background, networked forms of governance remain associated with the negative private sector influence over urban development and corruption.

In summary, the aforementioned factors contribute to more participation and higher levels of integration among actors from different sectors in Birmingham than in the other two cases. Frankfurt's network allows for participation (large network size), but integration among network actors is considerably lower – especially among the public and market sectors. The integration of external stakeholders is the least developed in Budapest, but despite the unfavourable conditions some network-type interaction (mainly among municipal companies and departments) appears to be necessary or beneficial to manage and deliver initiatives in a context characterised by the dominance of the national level. Patterns of organisational fragmentation and the distribution of resources and authority are, to some extent, also reflected in the local rationales for low-carbon transitions.

The development of participative networks and network integration however also depends largely on first, the local interpretations of complexity and uncertainties, rather than an exogenous characterisation of the energy transition problem as 'complex'. This relates to the nature of local low-carbon ambitions, and the options available to the Local Authorities to achieve these. Second, familiarity and locally accumulated experience with participative/networked forms of governance has also been an important factor. Consequently, historical and cultural factors cannot be neglected when attempting to understand the role and potential of network forms of governance in governing urban sustainability transitions.

It is also important to note that networks are no panacea – higher levels of integration do not necessarily improve the governability of low-carbon transitions. One reason for this can be that in well-connected networks the opportunities for undertaking bridging/intermediary roles are lacking (as in the case of Birmingham). This observation is particularly relevant for the intermediation perspective on governing urban low-carbon transitions. It highlights the importance of considering the broader network structure in which intermediary organisations are embedded (and the contextual factors that influence these structural features), in addition to their dyadic links to other network actors.

Contextual drivers of emerging power relationships within networks

Power relationships among different actors, and groups of actors according to sectoral origin, have been explored through statistical analyses of actor centralities in the energy transition networks in each city. Power relationships therefore refer to network-internal hierarchies. More balanced power relationships within the network are more likely to allow for the network mode of governance to arise and dominate the interaction processes between network actors. The literature on urban sustainability transitions emphasises the benefits that network governance offers in dealing with the complexities and uncertainties which they see as being intrinsic to transitions towards low-carbon futures. According to the context analyses, the following characteristics have been found to be influencing power relationships within the networks:

1 The role and position of local-level actors in the governance of energy systems, and the openness of the network towards new actors to join to counteract the negative effects of societal and organisational fragmentation.

2 A clearly defined direction for sustainable energy transitions signalling credibility and contributing to decreasing the locally perceived complexity and uncertainty of energy transitions.
3 Steering and leadership by the Local Authorities, including continued presence in the network relating to changing societal dynamics and the shift from 'government' to 'governance'.

First, in terms of societal and organisational fragmentation, power relationships within the network are influenced by the role and position of local-level actors in the governance of energy systems (the state and nature of decentralisation) and by the potential for new (local) actors to join. In the UK, the organisation of energy production, distribution and supply is centralised, and the design of the market favours large corporations. Consequently, local-level actors are in a relatively weak position, and possibilities for new actors to join, and grow, are limited despite the traditionally strong third sector (voluntary and community organisations). These factors lead to a strong position of market sector actors in Birmingham's network compared with the other two cities. Energy systems in Germany and in Frankfurt are rather decentralised. Mainova, the municipal energy company, appears as one of the most influential actors within the network. The national framework for energy transitions (Energiewende) encourages new entrants (small companies, citizen co-operatives, investment banks) to enter the sector, contributing to a stronger role for local actors in the energy transition network in Frankfurt. In Hungary, energy production and distribution systems are to some extent decentralised, but operated by companies outside the reach of Budapest City Council with the exception of district heating. As a result of a lack of commitment to further decentralisation, opportunities for new players to enter the energy sector are very limited. Budapest's network is therefore dominated by local-level actors who are also related to the City Council.

Second, the capacity to manage complexities and uncertainties related to sustainable energy transitions was, to some extent, dependent on setting a clear direction and role for energy sector in carbon emissions reduction. Characteristics contributing to a 'clear direction' included tangible targets specific to the energy sector, alignment between subnational and national targets, and a division of roles and responsibilities among the different organisational levels. In Birmingham, the City Council set an overall carbon emissions reduction target which (at the time) exceeded the national-level commitments – but without specifying the energy sector's expected part within this overall target. In addition, the absence of clear roles for the national and local levels in managing energy transitions (due to lacking national and local energy strategies) also contributed to a lack of credibility over the delivery of emissions reduction goals, which translated into more horizontal relationships and difficulties with managing external partners in the network. Of the three cities, only Frankfurt set carbon emissions targets specific to the energy sector, which were also aligned with the national-level goals. However, the strong economy based on energy-intensive sectors acts as a barrier to reducing emissions. In addition, international corporations who do not depend on the local market appear to be difficult to engage with, resulting in a dominant view (locally as well as nationally) that the market sector are customers whose energy demand must be satisfied. However, their role in the energy transition network remains unclear, leading to a network-internal hierarchy between Local Authority bodies and externals in Frankfurt's network. Budapest, similarly to Birmingham, only set itself a carbon emissions reduction target which corresponds to the national-level commitments. Whilst the expected contribution of the energy sector remains unclear, the role of the local level in delivering energy transitions appears to be relatively weak. Consequently, Budapest City Council decided not to take up the challenge of governing low-carbon energy transitions. Instead, the focus remains on piecemeal responses which correspond to national political priorities and national or EU funding opportunities. The lack of strategic leadership contributes to difficulties and a lack of interest in engaging with external stakeholders in more formal, strategic ways.

Third, the position of the actors within the energy transitions networks have been found to be influenced by their history of involvement in collaborative initiatives relevant to the evolving low-carbon agendas. In Birmingham, frequent organisational restructuration internally as well as between the local and regional levels resulted in a periodic presence of Local Authority bodies in the low-carbon energy network. Integration among external stakeholders over time, in which for example the Sustainability Team was not able to participate, resulted in a comparatively strong role for long-term network participants (private companies, consultancies, but also some community organisations). This led to a situation in which options to undertake network coordination (due to less apparent network-internal hierarchies) or intermediation (due to the interconnectedness of the network) were both limited for the Sustainability Team to whom these tasks were assigned. In contrast, Frankfurt City Council maintained interest and involvement in the low-carbon transition agenda despite changes on the political level due to the continued influence of The Greens. The Energy Agency / Energiereferat started engagement with external partners on the implementation/project delivery level, and continued to remain involved in network interactions for 30 years. This led to better opportunities to undertake network coordination. The Agency also contributed to intermediation between policy-making and implementation/delivery. In Budapest, a weak form of network coordination and intermediation was undertaken by the Municipal City Planning Agency (BFVT) – which, however, operates relatively autonomously rather than a municipal department, and has no decision-making authority on its own. This situation limits its potential to steer network processes, as decision-making authority is assigned to other actors. These actors (Deputy Mayor, Project Management Department, etc.), despite being involved in the network, do not undertake steering roles.

Power relationships within the networks were thus mainly influenced by the role and approach taken by Local Authority bodies (including the regional level where it exists) towards energy transitions. Local Authorities' role within the network-internal hierarchies is particularly interesting because, as the local state in democratic settings, they possess the mandate, authority and powers to develop visions, strategies and pathways towards low-carbon urban futures. Despite decision authority being concentrated in the public sector, the developing energy transition networks have the potential to influence local policy, strategic plans and implementation (cf. Christopoulos, 2008). Understanding power relationships within the networks, and acknowledging the reasons of their existence, therefore becomes important to assess who gains (more) influence over whom through the network interactions – and by doing so, over low-carbon futures (cf. Avelino, 2009). Moreover, it can also provide information on the locally appropriate steering roles, and the kind of coordination and intermediation activities that can and should be undertaken, for organisations to whom the delivery of low-carbon ambitions are assigned. This 'influence' perspective, and its implications for local democracy, is however not captured well in the existing literature on urban sustainability transitions. This leads to difficulties with delineating which kind of networks can contribute to meeting sustainability targets and which don't, and at what expense (cf. Walker and Shove, 2007). Moreover, it also gives the impression that the choice of undertaking different types of steering roles is a free choice, without acknowledging the locally important constraints and the potential for unintended consequences to arise based on this 'chosen' role.

Governing energy transitions in European cities: Conclusions

Linking networks, context and impact

The research into Birmingham, Frankfurt and Budapest revealed that governance networks, which have a role in governing local sustainable energy transitions, have existed in all three cities for some time. It also highlighted substantial differences among the three networks in terms of integration and power relationships among network actors due to the influence of local contextual factors.

Carbon emissions reduction, however, has been fairly slow in all three cities, with Frankfurt's per capita emissions remaining at the highest level. Per capita emissions thus appear to correlate with economic output (here denoted in terms of GDP; see Table 6.1). It is likely that rather than local efforts, emissions reduction in all three cases has so far mostly been driven by the ongoing decarbonisation of national electricity infrastructures, and the tendency of restructuring urban economies towards less carbon-intensive sectors such as finance and services (cf. While, 2014).

The impact of emerging governance networks on actual carbon emissions reduction therefore remains weak, signalling difficulties with steering networks towards delivering outcomes in terms of emissions reduction. This implies the need for identifying and implementing targeted interventions based on context-relevant knowledge – to better adapt the generic concept of governance networks to locally specific constraints and enablers.

Table 6.1 Emissions, targets and economic output in the case study cities

	Birmingham	*Budapest*	*Frankfurt*
Target	60% by 2027 100% by 2050	40% by 2030 (80% by 2050)	(55% by 2030) 95% by 2050
Emissions reductions achieved (per capita)	33.7% (2015; 1990 baseline), equivalent to a rate of decarbonisation of appr. 1.35%/year	15% (2015; 2005 baseline) equivalent to a rate of decarbonisation of appr. 1.5%/year	32% (2017; 1990 baseline) equivalent to a rate of decarbonisation of appr. 1.2%/year
Per capita emissions	4.4 t (2014)	5.3 t (2015)	9.8 t (2013)
Economic output (GDP, EUR)	25,500	38,900	88,600

Sources: BCC, no date; FCC, 2015; BuCC, 2018).

In Birmingham, energy transition governance is characterised by a lack of integration among Local Authority departments and a high degree of dependence on the national level. This leads to difficulties with developing a unified position within the City Council on decarbonisation goals, decreasing the credibility of low-carbon ambitions. Other contributing factors include the lack of leadership and specific targets and direction regarding the expected role of the energy sector in overall emissions reduction:

> I think where we'll always have the biggest difficulties is … because we are so siloed, because … the council exists in silos, and actually our national government exists in silos … That way of working still persists, partly because Whitehall in central government is siloed, and partly because we have followed, because that's where our money comes from.
>
> (Interview 1.12)

In terms of network impact, this leads to difficulties with coordinating network processes and, consequently, converting strategies 'on paper' into implementation. Network forms of governance could, however, focus specifically on developing a common voice and approach towards low-carbon transitions within the local public sector – including both the City Council and the Combined Authority. Moreover, delineating the expected contribution of the energy sector to the overall carbon emissions reduction targets could help decreasing overlaps among memberships and responsibilities of the different decision-making arenas, potentially contributing to easing the coordination problems arising from the large, polycentric network characteristics.

In Budapest, the legacy of the decision-making culture dominant under the communist system, and the way in which the transition to the capitalist system has been managed, acts as a barrier to developing more collaborative, participative forms of governance:

> [P]eople believe that everyone is each other's competitor. That if there is money available, it must be spent as soon as possible, because they [Central Government or the European Union] may take it back tomorrow. These instincts are still very strong today and this situation makes it difficult [to collaborate].
>
> (Interview 3.04)

In this context, Budapest City Council (rather unsurprisingly) chose not to pursue engagement with external partners. However, this results in a lack of capacity and capability to coordinate low-carbon transitions beyond its immediate environment. Developing better engagement processes with citizens is particularly pressing: while all municipal documents refer to the necessity of facilitating awareness and behaviour change, so far no concrete action has been taken to address this issue.

Based on historical developments, a sense of 'independence' has developed in Frankfurt. Attitudes to energy transitions reflect this, as well as the influential environmental movements' push for local energy independence and the democratisation of energy supply:

> [W]ithin the city we had the free rights of trade against the kings … Germany was based on these free rights, and it is also now … [Energy transitions] will not be and … should not be organised in a central way, we should not be organised by big companies, but by hundreds or thousands of decentralised associations and energy producers.
>
> (Interview 2.03)

This progressive agenda, however, appears to alienate market sector actors, particularly those international companies whose business does not depend on the local market. In addition, the high energy demand generated by the market sector also acts as a lock-in mechanism counteracting Mainova's efforts to decarbonise its energy generation (Interview 2.02; Interview 2.03). Interventions, including collaborative and participative decision-making initiatives, therefore should specifically be aimed at resolving this impasse.

In summary, while progress with the decarbonisation of energy infrastructures remains fairly slow in all three cities, the locally relevant causes of underlying problems that contribute to path-dependent development trajectories are different in each case. Although an acknowledgement that historical and cultural factors matter has been developing in the literature on sustainability transitions, these tend to concentrate on individual perspectives (e.g. beliefs, social practices, choices; (Shove and Walker, 2007; Avelino, 2009; Hölscher et al., 2019)) rather than on organisations and organisational systems operating within multi-level governance contexts (cf. Ehnert et al., 2018; Elsner, Monstadt and Raven, 2019). Analyses of specifically aimed at uncovering locally relevant niche roles for organisational networks, based on an in-depth understanding of the impact contextual opportunities and constraints, are found wanting. A more nuanced understanding of the local problems and the contextual conditions that cause them can help identifying specific niche roles that network forms of governance can play in facilitating and implementing local low-carbon transitions.

Not all network ties are created equal: A discussion on the diversity of network processes

In addition to contextual differences, network relationships and processes also exhibit significant variance. Network settings must therefore be analysed as enabling and constraining structures – without assuming too much in advance of the kinds of processes to which networks structures can give rise (Lewis, 2011). Based on the diversity of network processes, different parts of the network may fulfil various functions and/or niche roles (cf. Torfing and Sørensen, 2014).

Different types of processes have been identified in the energy transition networks in the three cities, with a potential of playing different types of roles relative to overall decision-making for energy transitions. These include: simple information exchange; networking and professional relationship-building and resource pooling; expert advice and consultation; and co-production. Table 6.2 shows the distribution of different types of network processes among the three city cases.

Decision-making arenas driven predominantly by information exchange were found in all three cities. In Budapest, this function was the most dominant in the existing informal networks (e.g. producing Inventories of the State of the Environment). Networking, professional relationship-building and resource pooling was characteristic of networks with higher levels of integration among the public and market sector and community organisations. Specifically, it was found to be dominant in implementation-orientated decision-making arenas with high levels of professional membership (e.g. Birmingham Science City

Table 6.2 Emerging network processes in the different energy transition networks

Network function	*Birmingham*	*Budapest*	*Frankfurt*
Information exchange	✓	✓	✓
Networking, professional relationship-building and resource pooling	✓		✓
Expert advice and consultation ('passive' involvement)	✓	✓	✓
Co-production ('active' involvement)	✓		

Low Carbon Working Group, Environmental Forum / Umweltforum). Expert advice and consultation type processes were more characteristic of Frankfurt (Climate Protection Advisory Group / Klimaschutzbeirat) and Budapest (ISCO – although formalised), but also found in Birmingham. More active involvement of external stakeholders, moving towards co-production, was found in Birmingham only (Green Commission, Roundtables).

These findings highlight the importance of further research into the diversity that network processes can take, and the ways in which these can (or cannot) function in different contextual settings – this will enable the development of a more realistic view of the potential of network forms of governance to support sustainability transitions. For example, this study indicated that it may prove difficult to successfully implement co-production type initiatives in contexts where significant network-internal hierarchical relationships exist. Network processes should therefore be set up based on context-specific information and sound analysis. Context-specific information must also include the local interpretations of networked forms of governance. In Budapest, networks, and private sector involvement in public sector decision-making, have a negative connotation and are associated with corruption. However, in order to improve transparency and integration, formalising some already existing interaction patterns could provide benefits for local energy transitions. In a similar vein, better integration through networks could help overcoming internal fragmentation among City Council departments.

Such focus on the diversity of network processes, and ways in which these can best support transitions in different local/urban settings is currently still lacking from the literature on urban sustainability transitions. This requires analysing sustainability transition networks as embedded in the broader governance arrangements and mechanisms (locally relevant hybrid combinations of hierarchical and market mechanisms and horizontal network coordination) that are relevant to sustainability transitions (cf. Skelcher, Sullivan and Jeffares, 2013). The research presented in this book provided one option for developing such a framework and methodology to deepen our understanding of the role of networks in the governance of low-carbon transitions.

The role of the network concept in multi-stakeholder decision-making: Analytical tool or metaphor?

Much of the literature on network governance tends to emphasise the risks involved in mixing quantitative (network analysis) and qualitative methods, resulting in a 'profound lack of comparative and cross-national studies and a striking failure to combine qualitative and quantitative approaches' (Torfing and Sørensen, 2014, p. 341). The study of energy transition networks in Birmingham, Budapest and Frankfurt, however, demonstrated that a carefully designed mixed methods approach can successfully contribute to deepening our understanding of the context-dependence of the network concept in urban governance. The methodology deployed in this study thus represents an innovative approach to study multi-actor decision-making processes and their dynamics.

The divide between conceptualising governance the 'network' either as a useful analytical tool (and a phenomenon worthy of investigation in its own right) or rather as metaphor has been reinforced by a number of factors. These include, for example, diverging interpretations of network connections and interactions (networks of influence versus networks of authority) and network actors (individuals versus organisations) (Borgatti and Halgin, 2011; Bergé, Wanzenböck and Scherngell, 2017; Christopoulos, 2017). Quantitative network-analysis-based investigations are more likely to be used with networks of individuals, while organisational networks are most often studied via qualitative methods (Marcussen and Olsen, 2007). This is rooted in the origins of network analysis (SNA – social network analysis) which main focus is on social networks of individuals and associated influences on their behaviour (Bergé, Wanzenböck and Scherngell, 2017). The issue of 'authority', however, results in complications with using the network method with networks of organisations because 'although influence can be perceived as diffuse, decision-making is reserved for those endowed with decision authority and is therefore concentrated' (Christopoulos, 2008, p. 477).

More recently, interest started to develop in applying mixed methods in the study of governance networks. The argument is that different methods can answer different types of questions, offering the potential to develop a more complete understanding of the subject of the study. For example, Lewis (2011, p. 1229) argues that quantitative network analysis can provide 'the context to help interpret network data' by providing an 'outside view of network structure to add to an inside view of the content, quality and meaning of network ties'.

This research demonstrated that, mixed methods research offers the possibility of generating comparable data via network visualisation and analysis. It can thus inform the qualitative enquiry into better understanding the context-specificity of the role and potential of network forms of governance in contributing to sustainability transitions in different places. Despite the fact that network analysis in itself would have left many questions unanswered, it provided initial insights and highlighted similarities and differences among the three cases to be further investigated, and validated, through qualitative research. Due to a lack of comparative analyses and frameworks, it is rather likely that the analysis presented in this book is not exhaustive. However, further applications and subsequent iterations can help to refine and test the methodology in more diverse settings, and better understand its transferability (for example, beyond Europe) and potential shortcomings or limitations.

References

Avelino, F. (2009) 'Empowerment and the challenge of applying transition management to ongoing projects', *Policy Sciences*, 42(4), pp. 369–390. doi:10.1007/s11077-009-9102-6.

BCC (no date) Energy and carbon reduction. Birmingham City Council. Available at: www.birmingham.gov.uk/info/20015/environment/260/energy_and_carbon_reduction (Accessed: 29 February 2020).

Bergé, L. R., Wanzenböck, I. and Scherngell, T. (2017) 'Centrality of regions in R&D networks: A new measurement approach using the concept of bridging paths', *Regional Studies*, 51(8), pp. 1165–1178. doi:10.1080/00343404.2016.1269885.

Borgatti, S. P. and Halgin, D. S. (2011) 'On network theory', *Organization Science*, 22(5), pp. 1168–1181. doi:10.1287/orsc.1100.0641.

BuCC (2018) Budapest Klímastratégiája. Available at: https://budapest.hu/Documents/klimastrategia/Bp_Klimastrategiája_vegleges_KGYelfogadott.pdf (Accessed: 29 February 2020).

Christopoulos, D. C. (2008) 'The governance of networks: Heuristic or formal analysis? A reply to Rachel Parker', *Political Studies*, 56(2), pp. 475–481. doi:10.1111/j.1467-9248.2008.00733.x.

Christopoulos, D. C. (2017) 'Governance networks in politics', in B. Hollstein, W. Matiaske and K.-U. Schnapp (eds), *Networked Governance: New Research Perspectives*, pp. 55–70. Springer. doi:10.1007/978-3-319-50386-8_4.

Ehnert, F.*et al.* (2018) 'Urban sustainability transitions in a context of multi-level governance: A comparison of four European states', *Environmental Innovation and Societal Transitions*, 26, pp. 101–116. doi:10.1016/j.eist.2017.05.002.

Elsner, I., Monstadt, J. and Raven, R. (2019) 'Decarbonising Rotterdam?', *City*, 23(4–5), pp. 646–657. doi:10.1080/13604813.2019.1689735.

Faure, A. M. (1994) 'Some methodological problems in comparative politics', *Journal of Theoretical Politics*, 6(3), pp. 307–322. doi:10.1177/0951692894006003003.

FCC (2015) Masterplan 100% climate protection. Frankfurt am Main. Available at: https://frankfurt.de/-/media/frankfurtde/frankfurt-themen/

klima-und-energie/pdf/energiereferat-79a/masterplan-climate-protection—summarised-version.ashx.

Hantrais, L. (1999) 'Contextualization in cross-national comparative research', *International Journal of Social Research Methodology*, 2(2), pp. 93–108. doi:10.1080/136455799295078.

Hantrais, L. (2013) *Cross-National Research Methodology and Practice*, edited by L. Hantrais and S. Mangen. Routledge. doi:10.4324/9781315878683.

Hölscher, K.*et al.* (2019) 'Opening up the transition arena: An analysis of (dis) empowerment of civil society actors in transition management in cities', *Technological Forecasting and Social Change*, 145, pp. 176–185. doi:10.1016/j.techfore.2017.05.004.

Kjaer, A. M. (2009) 'Governance and the urban bureaucracy', in J. S. Davies and D. L. Imbroscio (eds), *Theories of Urban Politics*, pp. 137–152. SAGE.

Lewis, J. M. (2011) 'The future of network governance research: Strength in diversity and synthesis', *Public Administration*, 89(4), pp. 1221–1234. doi:10.1111/j.1467-9299.2010.01876.x.

Marcussen, M. and Olsen, H. P. (2007) 'Transcending analytical cliquishness with second-generation governance network analysis', in J. Torfing and M. Marcussen (eds), *Democratic Network Governance in Europe*, pp. 273–292. Palgrave Macmillan. doi:10.1057/9780230596283_14.

Pierre, J. (2005) 'Comparative urban governance: Uncovering complex causalities', *Urban Affairs Review*, 40(4), pp. 446–462. doi:10.1177/1078087404273442.

Pollitt, C. and Bouckaert, G. (2011) *Public Management Reform – A Comparative Analysis: New Public Management, Governance, and the Neo-Weberian State*, 3rd edn. Oxford University Press.

Shove, E. and Walker, G. (2007) 'CAUTION! Transitions ahead: Politics, practice, and sustainable transition management', *Environment and Planning A*, 39(4), pp. 763–770.

Skelcher, C., Sullivan, H. and Jeffares, S. (2013) *Hybrid Governance in European Cities: Neighbourhood, Migration and Democracy*. Springer.

Torfing, J. (2005) 'Governance network theory: Towards a second generation', *European Political Science*, 4(3), pp. 305–315. doi:10.1057/palgrave.eps.2210031.

Torfing, J. and Sørensen, E. (2014) 'The European debate on governance networks: Towards a new and viable paradigm?', *Policy and Society*, 33(4), pp. 329–344. doi:10.1016/j.polsoc.2014.10.003.

Walker, G. and Shove, E. (2007) 'Ambivalence, sustainability and the governance of socio-technical transitions', *Journal of Environmental Policy and Planning*, 9(3–4), pp. 213–225. doi:10.1080/15239080701622840.

While, A. (2014) 'Carbon regulation and low-carbon urban restructuring', in M. Hodson and S. Marvin (eds), *After Sustainable Cities?*, pp. 41–58. Routledge.

Appendix I: Details of primary data collection and analysis

In addition to secondary data (municipal and partner websites, publications, meeting minutes (publicly available or accessible through FOI requests) and previous studies), the analysis of contextual change factors in the three cities also built on primary data collected through a series of interviews with key informants. The purpose of collecting qualitative data was to validate the results of the network analyses through describing the context in which the networks emerged and operated.

An initial set of interviewees were identified via the Transition Cities project, municipal policy documents and the results of the network analyses. Further interviewees were added based on snowball sampling (see lists of interviewees below). The interviews were semi-structured with only open-ended questions asked. Interviews were conducted in English in Birmingham and in Frankfurt, and in Hungarian in Budapest. The interviews focused on questions relating to the role of energy transitions in the local sustainability agenda, the drivers and barriers of transitions, associated governance initiatives and processes, the respondent's involvement in these and the experiences gained, and implications for low-carbon energy transitions. These categories were used to code the transcripts produced from the interview recordings.

In a small number of cases the interview also contributed to complementing and correcting the network visualisations; however, this was not a central goal of the interviews. Consequently, network visualisations were not discussed directly during the interviews. Rather, interview questions focused on verifying the findings arising from the network analyses, and unpacking the contextual factors that influenced the network structures and processes. Interviews were thus mainly used to describe the development of the local sustainable energy agendas, the role of transition arenas in local policy making and implementation, the relationships among the different arenas, and the nature and content of relationships between network actors.

Lists of Interviews

Birmingham

Interview 1.01. Interviewee, Procurement Team, Birmingham City Council (2015). Transition Cities representative. Interview with T. Nochta on 29/10/2015. Birmingham.

Interview 1.02. Interviewee, Sustainability West Midlands (2015). Green Commission member. Interview with T. Nochta on 16/11/2015. Birmingham.

Interview 1.03. Interviewee, Sustainability Team, Birmingham City Council (2015). Interview with T. Nochta on 15/12/2015. Birmingham.

Interview 1.04. Interviewee, Sustainability Team, Sustainability Team, Birmingham City Council (2016). Interview with T. Nochta on 2016/01/06. Birmingham.

Interview 1.05. Interviewee, Climate Change and Environment, Birmingham City Council (2016). Interview with T. Nochta on 12/01/2016. Birmingham.

Interview 1.06. Interviewee, Planning and Regeneration, Birmingham City Council (2016). Green Commission member. Interview with T. Nochta on 19/02/2016. Birmingham.

Interview 1.07. Interviewee, Sustainability West Midlands (2016). Birmingham Science City Low Carbon Working Group Member. Interview with T. Nochta on 12/02/2016. Birmingham.

Interview 1.08. Interviewee, Localise West Midlands and Community Energy Birmingham (2016). Green Commission member. Interview with T. Nochta on 26/02/2016. Birmingham.

Interview 1.09. Interviewee, Energy Technologies Institute (ETI) (2016). Green Commission member. Interview with T. Nochta on 01/03/2016. Loughborough.

Interview 1.10. Interviewee, Birmingham Energy Institute, University of Birmingham (2016). Green Commission member. Interview with T. Nochta on 04/03/2016. Birmingham.

Interview 1.11. Interviewee, Veolia (2016). Green Commission member. Interview with T. Nochta on 11/04/2016. Birmingham.

Interview 1.12. Interviewee, Birmingham City Council and Community Energy Birmingham (2016). Interview with T. Nochta on 29/04/2016. Birmingham.

Frankfurt

Interview 2.01. Interviewee, Energiereferat, Frankfurt City Council (2016). Former Transition Cities representative. Interview with T. Nochta on 19/05/2016. Frankfurt.

Interview 2.02. Interviewee, Energiereferat, Frankfurt City Council (2016). Interview with T. Nochta on 09/06/2016. Frankfurt.

Interview 2.03. Interviewee, Energiereferat, Frankfurt City Council (2016). Interview with T. Nochta on 13/06/2016. Frankfurt.

Interview 2.04. Interviewee, Sonneninitiative e.V. (2016). Umweltforum and Klimaschutzbeirat member. Interview with T. Nochta on 21/06/2016. Frankfurt.

Interview 2.05. Interviewee, Provadis Hochschule, Industriepark Hoechst (2016). Klimaschutzbeirat member. Interview with T. Nochta on 22/06/2016. Frankfurt.

Interview 2.06. Interviewee, Umweltlernen e.V., Frankfurt City Council (2016). Interview with T. Nochta on 22/06/2016. Frankfurt.

Interview 2.07. Interviewee, Energiereferat, Frankfurt City Council (2016). Interview with T. Nochta on 24/06/2016. Frankfurt.

Interview 2.08. Interviewee, Energiepunkt e.V., Frankfurt City Council (2016). Interview via e-mail with T. Nochta on 15/07/2016. Frankfurt.

Interview 2.09. Interviewee, Energiereferat, Frankfurt City Council (2016). Transition Cities representative. Interview with T. Nochta on 05/07/2016. Frankfurt.

Budapest

Interview 3.01. Interviewee, Project Management, Budapest City Council (2016). Transition Cities Representative. Interview with T. Nochta on 17/08/2016. Budapest.

Interview 3.02. Interviewee, Environmental Protection, Budapest City Council (2016). Interview with T. Nochta on 01/09/2016. Budapest.

Interview 3.03. Interviewee, Energy and Water Management, Budapest City Council (2016). Interview with T. Nochta on 09/09/2016. Budapest.

Interview 3.04. Interviewee, Budapest Zoo (2016). Interview with T. Nochta on 09/09/2016. Budapest.

Interview 3.05. Interviewee, Budapest City Council (2016). Interview with T. Nochta on 20/09/2016. Budapest.

Interview 3.06. Interviewee, Óbuda (District III) District Council (2016). Interview with T. Nochta on 20/09/2016. Budapest.

Interview 3.07. Interviewee, FOTAV Zrt (2016). Interview with T. Nochta on 21/09/2016. Budapest.

Interview 3.08. Interviewee, BFVT Zrt (2016). Interview with T. Nochta on 21/09/2016. Budapest.

Interview 3.09. Interviewee, GRID Consulting (2016). Interview with T. Nochta on 23/09/2016. Budapest.

Interview 3.10. Interviewee, Energiaklub Climate Policy Institute (2016). Interview with T. Nochta on 23/09/2016. Budapest.

Interview 3.11. Interviewee, FKF Zrt (2016). Interview with T. Nochta on 30/09/2016. Budapest.

Interview 3.12. Interviewee, Environmental Protection and Quality Control, BERT Zrt (2016). Interview with T. Nochta on 02/10/2016. Budapest.

Appendix II: Network analysis: Actor centrality scores (core groups)

Table II.1 Degree centrality

Rank	*Organisation*	*Degree**	*Organisation*	*Degree (core)**
		Birmingham		
1	Aston University	0.76	Aston University	1.00
2	University of Birmingham	0.76	British Gas	1.00
3	British Gas	0.69	University of Birmingham	0.96
4	Innovation Birmingham – Climate KIC	0.67	Midlands Environmental Business Company	0.94
5	Sustainability West Midlands	0.66	BCC Cabinet	0.92
6	West Midlands Integrated Transport Authority	0.66	West Midlands Integrated Transport Authority	0.92
7	Cenex	0.65	Sustainability West Midlands	0.92
8	Encraft	0.64	Innovation Birmingham – Climate KIC	0.90
9	Midlands Environmental Business Company	0.63	Localise West Midlands	0.90
10	BCC Sustainability	0.61	BCC Sustainability	0.88
11	Sustainable Action Housing Partnership	0.61	BVSC Third Sector Assembly	0.81
12	Coventry City Council	0.59	Cenex	0.81
13	Marches Energy Agency	0.59	Community Energy Birmingham	0.81
14	Severn Trent Water	0.59	Severn Trent Water	0.81
15	Localise West Midlands	0.57	Solihull Metropolitan Borough Council	0.81
16	Solihull Metropolitan Borough Council	0.55	ENGIE	0.79

Rank	*Organisation*	*Degree**	*Organisation*	*Degree (core)**
17	BCC Cabinet	0.54	GBS LEP	0.79
18	Staffordshire County Council	0.53	Amey plc	0.77
19	GBS LEP	0.52	Atkins	0.77
20	Environment Agency	0.50	Encraft	0.77
21	ENGIE	0.47	Sustainable Action Housing Partnership	0.77
22	Amey plc	0.46	Coventry City Council	0.75
23	National Farmers Union West Midlands	0.46	Marches Energy Agency	0.75
24	BCC Digital Birmingham	0.46	Energy Saving Trust	0.69
25	Veolia	0.45	Environment Agency	0.69
26	Atkins	0.41	Marketing Birmingham	0.69
27	Briar Associates	0.41	Staffordshire County Council	0.69
28	Caplor Energy	0.41	BCC Digital Birmingham	0.67
29	Caterpillar (UK) Ltd	0.41	International Synergies	0.67
30	Lafarge Tarmac	0.41	BCC Planning and Regeneration	0.65
31	Worcestershire County Council	0.41	Veolia	0.65
32	BCC Planning and Regeneration	0.37	BCC Transport	0.60
33	Energy Saving Trust	0.37	European Bioenergy Research Institute	0.60
34	Marketing Birmingham	0.36	Northfield Ecocentre	0.60

(*continued*)

Table II.1 (continued)

Rank	*Organisation*	*Degree**	*Organisation*	*Degree (core)**
35	BVSC Third Sector Assembly	0.35	E-ON	0.58
36	Community Energy Birmingham	0.35	Greater Birmingham Chambers of Commerce	0.58
37	European Bioenergy Research Institute	0.35	National Farmers Union West Midlands	0.58
38	Skanska	0.34	Skanska	0.56
39	Centro	0.33	Briar Associates	0.54
40	Greater Birmingham Chambers of Commerce	0.31	Caplor Energy	0.54
41	International Synergies	0.30	Caterpillar (UK) Ltd	0.54
42	BCC Transport	0.29	Centro	0.54
43	Northfield Ecocentre	0.28	Lafarge Tarmac	0.54
44	E-ON	0.27	Worcestershire County Council	0.54
45	Wyre Forest District Council	0.25	Birmingham & Black Country Local Nature Partnership	0.52
46	East Staffordshire Borough Council	0.23	Wyre Forest District Council	0.50
47	Birmingham & Black Country Local Nature Partnership	0.22	East Staffordshire Borough Council	0.40
48	ACIVICO	0.20	ACIVICO	0.35
49	Birmingham Metropolitan College	0.18	Birmingham Metropolitan College	0.35

Rank	*Organisation*	*Degree**	*Organisation*	*Degree (core)**
		Frankfurt		
1	Mainova AG	0.80	Regionalverband FRM	0.98
2	Sonneninitiative e.V.	0.76	Energiereferat, Stadt Frankfurt	0.93
3	FRAPORT AG, Airport	0.74	Mainova AG	0.93
4	ABGnova GmbH	0.74	InfraServ GmbH & Co. Hochst KG	0.85
5	Frankfurt University of Applied Sciences	0.74	IHK Chambers of Commerce and Industry	0.83
6	Landessportbund Hessen e.V.	0.67	Handwerkskammer FRM	0.80
7	Stadt Maintal	0.67	Stadt Friedrichsdorf	0.80
8	Burger AG fur Nachhaltiges Wirtschaften FRM	0.67	Institut fur Sozial-ekologische Forschung ISOE GmbH	0.78
9	Stadtmobil Rhein-Main GmbH	0.67	FRAPORT AG, Airport	0.76
10	Umweltamt, Stadt Frankfurt	0.65	ABGnova GmbH	0.74
11	Frankfurter Sparkasse AG	0.61	Burger AG fur Nachhaltiges Wirtschaften FRM	0.74
12	GLS Gemeinschaftsbank eG	0.61	Frankfurt University of Applied Sciences	0.74
13	SolarInvest Main-Taunus eG	0.61	Hochschule Darmstadt	0.74
14	Regionalverband FRM	0.43	Stadt Maintal	0.70
15	Energiereferat, Stadt Frankfurt	0.38	Cabinet, Stadt Frankfurt	0.67
16	IHK Chambers of Commerce and Industry	0.32	Provadis School	0.67

(continued)

Table II.1 (*continued*)

Rank	*Organisation*	*Degree**	*Organisation*	*Degree (core)**
17	InfraServ GmbH & Co. Hochst KG	0.31	DEHOGA KV Frankfurt e.V.	0.65
18	Institut fur Sozial-ekologische Forschung ISOE GmbH	0.31	FES GmbH	0.65
19	Handwerkskammer FRM	0.27	Johann Wolfgang Goethe-Universitat	0.65
20	Stadt Hattersheim	0.26	BUND Landesverband Hessen e.V.	0.63
21	Johann Wolfgang Goethe-Universitat	0.24	Evangelische Kirche Hessen&Nassau	0.63
22	Evangelische Kirche Hessen&Nassau	0.24	House of Logistics & Mobility (HOLM) GmbH	0.63
23	BUND Landesverband Hessen e.V.	0.24	Verkehrsclub Deutschland e.V.	0.63
24	Cabinet, Stadt Frankfurt	0.23	Energiepunkt FrankfurtRheinMain e.V.	0.61
25	Provadis School	0.23	schneider+schumacher Planungsgesellschaft mbH	0.61
26	Stadt Neu-Isenburg	0.22	Verbraucherzentrale Hessen e.V.	0.61
27	Hochschule Darmstadt	0.21	FED, Stadt Frankfurt	0.59
28	FED, Stadt Frankfurt	0.20	Kreislandwirt Frankfurt	0.59
29	DEHOGA KV Frankfurt e.V.	0.20	Stadt Offenbach am Main	0.57
30	FES GmbH	0.20	Umweltamt, Stadt Frankfurt	0.54
31	Verkehrsclub Deutschland e.V.	0.20	Landessportbund Hessen e.V.	0.52
32	House of Logistics & Mobility (HOLM) GmbH	0.20	Stadt Hattersheim	0.52

Rank	*Organisation*	*Degree**	*Organisation*	*Degree (core)**
33	Energiepunkt FrankfurtRheinMain e.V.	0.20	Stadt Neu-Isenburg	0.52
34	Verbraucherzentrale Hessen e.V.	0.20	Suwag Energie AG	0.52
35	schneider+schumacher Planungsgesellschaft mbH	0.20	SolarInvest Main-Taunus eG	0.50
36	Stadt Offenbach am Main	0.19	Stadt Russelsheim	0.50
37	Stadt Friedrichsdorf	0.19	Planung und Wohnen, Stadt Frankfurt	0.46
38	Siemens AG	0.17	Wirtschaft, Sport, Sicherheit und Feuerwehr, Stadt Frankfurt	0.46
39	Kreislandwirt Frankfurt	0.17	Frankfurter Sparkasse AG	0.39
40	Planung und Wohnen, Stadt Frankfurt	0.16	GLS Gemeinschaftsbank eG	0.39
41	Wirtschaft, Sport, Sicherheit und Feuerwehr, Stadt Frankfurt	0.16	Hessisches Ministerium fur Wirtschaft, Energie, Verkehr und Landesentwicklung	0.39
42	Climate Alliance	0.12	Siemens AG	0.39
43	Stadt Russelsheim	0.12	Sonneninitiative e.V.	0.39
44	Industrial Union of Mining, Chemical and Energy Industries, IG BCE	0.11	Industrial Union of Mining, Chemical and Energy Industries, IG BCE	0.37
45	VhU Association of Hessian Entrepreneurs	0.11	VhU Association of Hessian Entrepreneurs	0.37
46	Hessisches Ministerium fur Wirtschaft, Energie, Verkehr und Landesentwicklung	0.11	Climate Alliance	0.35
47	Suwag Energie AG	0.11	Stadtmobil Rhein-Main GmbH	0.35

(*continued*)

Table II.1 (continued)

Rank	*Organisation*	*Degree**	*Organisation*	*Degree (core)**
		Budapest		
1	BFVT Kft	1.00	BFVT Kft	1.00
2	BuCC Deputy Mayor/Advisor	0.85	BuCC Deputy Mayor/Advisor	1.00
3	BuCC Planning	0.77	BuCC Project Management	1.00
4	BuCC Project Management	0.77	BuCC Planning	1.00
5	BuCC Commission	0.46	FV Zrt.	0.92
6	BGYH Zrt.	0.65	BKK Zrt.	0.92
7	BKK Zrt.	0.65	BKV Zrt.	0.92
8	BKV Zrt.	0.65	BGYH Zrt.	0.92
9	BuCC Environment	0.65	FKF Zrt.	0.92
10	FKF Zrt.	0.65	FOKETUSZ Kft.	0.92
11	FOKETUSZ Kft.	0.65	FOTAV Zrt.	0.92
12	FOTAV Zrt.	0.65	FTSZV Kft.	0.92
13	FTSZV Kft.	0.65	BuCC Environment	0.92
14	FV Zrt.	0.65	BuCC Commission	0.31

Table II.2 Betweenness centrality

Rank	*Organisation*	*Betweenness**	*Organisation*	*Betweenness (core)**
		Birmingham		
1	Solihull Metropolitan Borough Council	0.0549	Aston University	0.0203
2	Encraft	0.0528	British Gas	0.0203
3	GBS LEP	0.0521	Midlands Environmental Business Company	0.0169
4	British Gas	0.0429	University of Birmingham	0.0155
5	University of Birmingham	0.0396	BCC Cabinet	0.0136
6	National Farmers Union West Midlands	0.0386	Localise West Midlands	0.0134
7	Aston University	0.0352	Innovation Birmingham – Climate KIC	0.0133
8	BCC Cabinet	0.0326	Sustainability West Midlands	0.0122
9	BCC Sustainability	0.0326	West Midlands Integrated Transport Authority	0.0122
10	Midlands Environmental Business Company	0.0322	BCC Sustainability	0.0118
11	Cenex	0.0239	BVSC Third Sector Assembly	0.0106
12	West Midlands Integrated Transport Authority	0.0238	Community Energy Birmingham	0.0106
13	Severn Trent Water	0.0216	Solihull Metropolitan Borough Council	0.0103
14	Sustainable Action Housing Partnership	0.0210	Severn Trent Water	0.0095
15	Innovation Birmingham – Climate KIC	0.0209	GBS LEP	0.0081
16	Sustainability West Midlands	0.0173	Sustainable Action Housing Partnership	0.0075

(*continued*)

Table II.2 (continued)

Rank	*Organisation*	*Between-ness**	*Organisation*	*Between-ness (core)**
17	Localise West Midlands	0.0137	Encraft	0.0072
18	BCC Digital Birmingham	0.0131	Cenex	0.0069
19	Amey plc	0.0119	ENGIE	0.0067
20	Environment Agency	0.0117	Amey plc	0.0066
21	Coventry City Council	0.0114	BCC Planning and Regeneration	0.0056
22	Marches Energy Agency	0.0114	Atkins	0.0052
23	ENGIE	0.0113	Veolia	0.0051
24	BCC Planning and Regeneration	0.0098	Coventry City Council	0.0051
25	Atkins	0.0080	Marches Energy Agency	0.0051
26	Energy Saving Trust	0.0078	Energy Saving Trust	0.0050
27	Staffordshire County Council	0.0076	International Synergies	0.0049
28	Birmingham Metropolitan College	0.0076	Marketing Birmingham	0.0048
29	Veolia	0.0073	Environment Agency	0.0046
30	Greater Birmingham Chambers of Commerce	0.0062	Staffordshire County Council	0.0040
31	BCC Transport	0.0045	National Farmers Union West Midlands	0.0038
32	E-ON	0.0042	BCC Digital Birmingham	0.0035
33	Skanska	0.0042	E-ON	0.0027

Rank	*Organisation*	*Between-ness**	*Organisation*	*Between-ness (core)**
34	BVSC Third Sector Assembly	0.0041	Northfield Ecocentre	0.0024
35	Community Energy Birmingham	0.0041	Greater Birmingham Chambers of Commerce	0.0023
36	Marketing Birmingham	0.0037	European Bioenergy Research Institute	0.0023
37	Wyre Forest District Council	0.0037	BCC Transport	0.0022
38	European Bioenergy Research Institute	0.0037	Skanska	0.0016
39	East Staffordshire Borough Council	0.0035	Centro	0.0011
40	Centro	0.0034	Wyre Forest District Council	0.0010
41	International Synergies	0.0031	East Staffordshire Borough Council	0.0007
42	Northfield Ecocentre	0.0029	Birmingham & Black Country Local Nature Partnership	0.0006
43	Briar Associates	0.0019	Briar Associates	0.0005
44	Caplor Energy	0.0019	Caplor Energy	0.0005
45	Caterpillar (UK) Ltd	0.0019	Caterpillar (UK) Ltd	0.0005
46	Lafarge Tarmac	0.0019	Lafarge Tarmac	0.0005
47	Worcestershire County Council	0.0019	Worcestershire County Council	0.0005
48	Birmingham & Black Country Local Nature Partnership	0.0012	Birmingham Metropolitan College	0.0003
49	ACIVICO	0.0011	ACIVICO	0.0003

(*continued*)

Table II.2 (continued)

Rank	*Organisation*	*Betweenness**	*Organisation*	*Betweenness(core)**
		Frankfurt		
1	Mainova AG	0.0886	Regionalverband FRM	0.0385
2	Sonneninitiative e.V.	0.0648	Mainova AG	0.0354
3	Frankfurt University of Applied Sciences	0.0553	Energiereferat, Stadt Frankfurt	0.0328
4	FRAPORT AG, Airport	0.0487	InfraServ GmbH & Co. Hochst KG	0.0216
5	ABGnova GmbH	0.0475	Sonneninitiative e.V.	0.0204
6	Regionalverband FRM	0.0401	Handwerkskammer FRM	0.0173
7	Stadtmobil Rhein-Main GmbH	0.0297	IHK Chambers of Commerce and Industry	0.0170
8	Landessportbund Hessen e.V.	0.0290	Hochschule Darmstadt	0.0153
9	Stadt Maintal	0.0290	BUND Landesverband Hessen e.V.	0.0143
10	Umweltamt, Stadt Frankfurt	0.0272	Frankfurt University of Applied Sciences	0.0133
11	Burger AG fur Nachhaltiges Wirtschaften FRM	0.0226	Stadt Hattersheim	0.0132
12	Energiereferat, Stadt Frankfurt	0.0199	Evangelische Kirche Hessen&Nassau	0.0127
13	IHK Chambers of Commerce and Industry	0.0167	FRAPORT AG, Airport	0.0110
14	InfraServ GmbH & Co. Hochst KG	0.0166	ABGnova GmbH	0.0102
15	Handwerkskammer FRM	0.0121	Institut fur Sozial-ekologische Forschung ISOE GmbH	0.0101
16	Frankfurter Sparkasse AG	0.0109	Johann Wolfgang Goethe-Universitat	0.0082

Rank	*Organisation*	*Betweenness**	*Organisation*	*Betweenness(core)**
17	GLS Gemeinschaftsbank eG	0.0109	FED, Stadt Frankfurt	0.0082
18	SolarInvest Main-Taunus eG	0.0109	Umweltamt, Stadt Frankfurt	0.0077
19	Stadt Hattersheim	0.0088	Cabinet, Stadt Frankfurt	0.0070
20	Institut fur Sozial-ekologische Forschung ISOE GmbH	0.0084	Provadis School	0.0070
21	Johann Wolfgang Goethe-Universitat	0.0083	Stadt Friedrichsdorf	0.0068
22	Evangelische Kirche Hessen&Nassau	0.0078	Burger AG fur Nachhaltiges Wirtschaften FRM	0.0065
23	Cabinet, Stadt Frankfurt	0.0078	Siemens AG	0.0057
24	Provadis School	0.0078	Landessportbund Hessen e.V.	0.0044
25	FED, Stadt Frankfurt	0.0056	Stadt Maintal	0.0044
26	Hochschule Darmstadt	0.0042	Stadtmobil Rhein-Main GmbH	0.0043
27	Stadt Neu-Isenburg	0.0042	DEHOGA KV Frankfurt e.V.	0.0042
28	BUND Landesverband Hessen e.V.	0.0042	FES GmbH	0.0042
29	Siemens AG	0.0039	Stadt Neu-Isenburg	0.0035
30	DEHOGA KV Frankfurt e.V.	0.0032	House of Logistics & Mobility (HOLM) GmbH	0.0033
31	FES GmbH	0.0032	Verkehrsclub Deutschland e.V.	0.0033
32	Planung und Wohnen, Stadt Frankfurt	0.0031	Energiepunkt FrankfurtRheinMain e.V.	0.0032
33	Wirtschaft, Sport, Sicherheit und Feuerwehr, Stadt Frankfurt	0.0031	schneider+schumacher Planungsgesellschaft mbH	0.0032
34	Stadt Friedrichsdorf	0.0030	Verbraucherzentrale Hessen e.V.	0.0032

(*continued*)

Table II.2 (continued)

Rank	*Organisation*	*Between-ness**	*Organisation*	*Betweenn-ess(core)**
35	Stadt Offenbach am Main	0.0018	Stadt Offenbach am Main	0.0026
36	House of Logistics & Mobility (HOLM) GmbH	0.0014	Kreislandwirt Frankfurt	0.0024
37	Verkehrsclub Deutschland e.V.	0.0014	Hessisches Ministerium fur Wirtschaft, Energie, Verkehr und Landesentwicklung	0.0024
38	Energiepunkt FrankfurtRheinMain e.V.	0.0013	Suwag Energie AG	0.0024
39	schneider+schumacher Planungsge-sellschaft mbH	0.0013	Climate Alliance	0.0023
40	Verbraucherzentrale Hessen e.V.	0.0013	Stadt Russelsheim	0.0023
41	Industrial Union of Mining, Chemical and Energy Industries, IG BCE	0.0009	Frankfurter Sparkasse AG	0.0022
42	VhU Association of Hessian Entrepreneurs	0.0009	GLS Gemeinschaftsbank eG	0.0022
43	Climate Alliance	0.0008	SolarInvest Main-Taunus eG	0.0022
44	Stadt Russelsheim	0.0008	Planung und Wohnen, Stadt Frankfurt	0.0019
45	Kreislandwirt Frankfurt	0.0007	Wirtschaft, Sport, Sicherheit und Feuerwehr, Stadt Frankfurt	0.0019
46	Hessisches Ministerium fur Wirtschaft, Energie, Verkehr und Landesentwicklung	0.0003	Industrial Union of Mining, Chemical and Energy Industries, IG BCE	0.0009
47	Suwag Energie AG	0.0003	VhU Association of Hessian Entrepreneurs	0.0009

Rank	*Organisation*	*Betwen-ness**	*Organisation*	*Betweenn-ess(core)**
		Budapest		
1	BFVT Kft	0.2167	BFVT Kft	0.0288
2	BuCC Deputy Mayor/Advisor	0.1010	BuCC Deputy Mayor/Advisor	0.0288
3	BuCC Project Management	0.0538	BuCC Project Management	0.0288
4	BuCC Planning	0.0538	BuCC Planning	0.0288
5	BuCC Commission	0.0164	FV Zrt.	0.0000
6	FV Zrt.	0.0049	BKK Zrt.	0.0000
7	BKK Zrt.	0.0049	BKV Zrt.	0.0000
8	BKV Zrt.	0.0049	BGYH Zrt.	0.0000
9	BGYH Zrt.	0.0049	FKF Zrt.	0.0000
10	FKF Zrt.	0.0049	FOKETUSZ Kft.	0.0000
11	FOKETUSZ Kft.	0.0049	FOTAV Zrt.	0.0000
12	FOTAV Zrt.	0.0049	FTSZV Kft.	0.0000
13	FTSZV Kft.	0.0049	BuCC Environment	0.0000
14	BuCC Environment	0.0049	BuCC Commission	0.0000

Table II.3 Closeness centrality

Rank	*Organisation*	*Closeness**	*Organisation*	*Closeness (core)**
		Birmingham		
1	Aston University	0.80	Aston University	1.00
2	University of Birmingham	0.80	British Gas	1.00
3	British Gas	0.77	University of Birmingham	0.96
4	Innovation Birmingham – Climate KIC	0.75	Midlands Environmental Business Company	0.94
5	Sustainability West Midlands	0.75	BCC Cabinet	0.92
6	West Midlands Integrated Transport Authority	0.75	Sustainability West Midlands	0.92
7	Cenex	0.74	West Midlands Integrated Transport Authority	0.92
8	Encraft	0.74	Innovation Birmingham – Climate KIC	0.91
9	Midlands Environmental Business Company	0.73	Localise West Midlands	0.91
10	BCC Sustainability	0.72	BCC Sustainability	0.89
11	Sustainable Action Housing Partnership	0.72	BVSC Third Sector Assembly	0.84
12	Coventry City Council	0.71	Cenex	0.84
13	Marches Energy Agency	0.71	Community Energy Birmingham	0.84
14	Severn Trent Water	0.71	Severn Trent Water	0.84
15	Localise West Midlands	0.70	Solihull Metropolitan Borough Council	0.84
16	Solihull Metropolitan Borough Council	0.69	ENGIE	0.83

Rank	*Organisation*	*Closeness**	*Organisation*	*Closeness (core)**
17	BCC Cabinet	0.68	GBS LEP	0.83
18	Staffordshire County Council	0.68	Amey plc	0.81
19	GBS LEP	0.68	Atkins	0.81
20	Environment Agency	0.67	Encraft	0.81
21	ENGIE	0.65	Sustainable Action Housing Partnership	0.81
22	Amey plc	0.65	Coventry City Council	0.80
23	National Farmers Union West Midlands	0.65	Marches Energy Agency	0.80
24	BCC Digital Birmingham	0.65	Energy Saving Trust	0.76
25	Veolia	0.64	Environment Agency	0.76
26	Atkins	0.63	Marketing Birmingham	0.76
27	Briar Associates	0.63	Staffordshire County Council	0.76
28	Caplor Energy	0.63	BCC Digital Birmingham	0.75
29	Caterpillar (UK) Ltd	0.63	International Synergies	0.75
30	Lafarge Tarmac	0.63	BCC Planning and Regeneration	0.74
31	Worcestershire County Council	0.63	Veolia	0.74
32	BCC Planning and Regeneration	0.61	BCC Transport	0.72
33	Energy Saving Trust	0.61	European Bioenergy Research Institute	0.72
34	Marketing Birmingham	0.61	Northfield Ecocentre	0.72
35	BVSC Third Sector Assembly	0.60	E-ON	0.71

(*continued*)

Table II.3 (continued)

Rank	*Organisation*	*Closeness**	*Organisation*	*Closeness (core)**
36	Community Energy Birmingham	0.60	Greater Birmingham Chambers of Commerce	0.71
37	European Bioenergy Research Institute	0.60	National Farmers Union West Midlands	0.71
38	Skanska	0.60	Skanska	0.70
39	Centro	0.60	Briar Associates	0.69
40	Greater Birmingham Chambers of Commerce	0.59	Caplor Energy	0.69
41	International Synergies	0.59	Caterpillar (UK) Ltd	0.69
42	BCC Transport	0.59	Centro	0.69
43	Northfield Ecocentre	0.58	Lafarge Tarmac	0.69
44	E-ON	0.58	Worcestershire County Council	0.69
45	Wyre Forest District Council	0.57	Birmingham & Black Country Local Nature Partnership	0.68
46	East Staffordshire Borough Council	0.57	Wyre Forest District Council	0.67
47	Birmingham & Black Country Local Nature Partnership	0.56	East Staffordshire Borough Council	0.62
48	Birmingham Metropolitan College	0.55	ACIVICO	0.61
49	ACIVICO	0.54	Birmingham Metropolitan College	0.61

Rank	*Organisation*	*Closeness**	*Organisation*	*Closeness (core)**
		Frankfurt		
1	Mainova AG	0.83	Regionalverband FRM	0.98
2	Sonneninitiative e.V.	0.81	Energiereferat, Stadt Frankfurt	0.94
3	ABGnova GmbH	0.80	Mainova AG	0.94
4	FRAPORT AG, Airport	0.80	InfraServ GmbH & Co. Hochst KG	0.87
5	Frankfurt University of Applied Sciences	0.79	IHK Chambers of Commerce and Industry	0.85
6	Burger AG fur Nachhaltiges Wirtschaften FRM	0.75	Handwerkskammer FRM	0.84
7	Landessportbund Hessen e.V.	0.75	Sonneninitiative e.V.	0.84
8	Stadt Maintal	0.75	Institut fur Sozial-ekologische Forschung ISOE GmbH	0.82
9	Stadtmobil Rhein-Main GmbH	0.75	FRAPORT AG, Airport	0.81
10	Umweltamt, Stadt Frankfurt	0.74	ABGnova GmbH	0.79
11	Frankfurter Sparkasse AG	0.72	BUND Landesverband Hessen e.V.	0.79
12	GLS Gemeinschaftsbank eG	0.72	Frankfurt University of Applied Sciences	0.79
13	SolarInvest Main-Taunus eG	0.72	Hochschule Darmstadt	0.79
14	Regionalverband FRM	0.64	Stadt Hattersheim	0.77
15	Energiereferat, Stadt Frankfurt	0.62	Cabinet, Stadt Frankfurt	0.75
16	IHK Chambers of Commerce and Industry	0.60	Provadis School	0.75

(*continued*)

Table II.3 (continued)

Rank	*Organisation*	*Closeness**	*Organisation*	*Closeness (core)**
17	InfraServ GmbH & Co. Hochst KG	0.59	DEHOGA KV Frankfurt e.V.	0.74
18	Institut fur Sozial-ekologische Forschung ISOE GmbH	0.59	FES GmbH	0.74
19	Handwerkskammer FRM	0.58	Johann Wolfgang Goethe-Universitat	0.74
20	Stadt Hattersheim	0.58	Burger AG fur Nachhaltiges Wirtschaften FRM	0.73
21	Johann Wolfgang Goethe-Universitat	0.57	Evangelische Kirche Hessen&Nassau	0.73
22	BUND Landesverband Hessen e.V.	0.57	House of Logistics & Mobility (HOLM) GmbH	0.73
23	Evangelische Kirche Hessen&Nassau	0.57	Verkehrsclub Deutschland e.V.	0.73
24	Cabinet, Stadt Frankfurt	0.57	Energiepunkt FrankfurtRheinMain e.V.	0.72
25	Provadis School	0.57	schneider+schumacher Planungsgesellschaft mbH	0.72
26	Stadt Neu-Isenburg	0.56	Verbraucherzentrale Hessen e.V.	0.72
27	Hochschule Darmstadt	0.56	FED, Stadt Frankfurt	0.71
28	DEHOGA KV Frankfurt e.V.	0.55	Kreislandwirt Frankfurt	0.71
29	Energiepunkt FrankfurtRheinMain e.V.	0.55	Stadt Neu-Isenburg	0.70
30	FED, Stadt Frankfurt	0.55	Umweltamt, Stadt Frankfurt	0.69
31	FES GmbH	0.55	Landessportbund Hessen e.V.	0.68
32	House of Logistics & Mobility (HOLM) GmbH	0.55	Stadt Friedrichsdorf	0.68

Rank	*Organisation*	*Closeness**	*Organisation*	*Closeness (core)**
33	schneider+schumacher Planungsgesellschaft mbH	0.55	Stadt Maintal	0.68
34	Verbraucherzentrale Hessen e.V.	0.55	Stadtmobil Rhein-Main GmbH	0.68
35	Verkehrsclub Deutschland e.V.	0.55	Siemens AG	0.67
36	Stadt Offenbach am Main	0.55	Stadt Offenbach am Main	0.67
37	Stadt Friedrichsdorf	0.55	Planung und Wohnen, Stadt Frankfurt	0.65
38	Kreislandwirt Frankfurt	0.55	Wirtschaft, Sport, Sicherheit und Feuerwehr, Stadt Frankfurt	0.65
39	Siemens AG	0.55	Frankfurter Sparkasse AG	0.62
40	Planung und Wohnen, Stadt Frankfurt	0.54	GLS Gemeinschaftsbank eG	0.62
41	Wirtschaft, Sport, Sicherheit und Feuerwehr, Stadt Frankfurt	0.54	Hessisches Ministerium fur Wirtschaft, Energie, Verkehr und Landesentwicklung	0.62
42	Climate Alliance	0.53	SolarInvest Main-Taunus eG	0.62
43	Stadt Russelsheim	0.53	Suwag Energie AG	0.62
44	Hessisches Ministerium fur Wirtschaft, Energie, Verkehr und Landesentwicklung	0.53	Industrial Union of Mining, Chemical and Energy Industries, IG BCE	0.61
45	Industrial Union of Mining, Chemical and Energy Industries, IG BCE	0.53	VhU Association of Hessian Entrepreneurs	0.61
46	Suwag Energie AG	0.53	Climate Alliance	0.61
47	VhU Association of Hessian Entrepreneurs	0.53	Stadt Russelsheim	0.61

(*continued*)

Table II.3 (continued)

Rank	*Organisation*	*Closeness**	*Organisation*	*Closeness (core)**
		Budapest		
1	BFVT Kft	1.00	BFVT Kft	1.00
2	BuCC Deputy Mayor/Advisor	0.87	BuCC Deputy Mayor/Advisor	1.00
3	BuCC Project Management	0.81	BuCC Project Management	1.00
4	BuCC Planning	0.81	BuCC Planning	1.00
5	FV Zrt.	0.74	FV Zrt.	0.93
6	BKK Zrt.	0.74	BKK Zrt.	0.93
7	BKV Zrt.	0.74	BKV Zrt.	0.93
8	BGYH Zrt.	0.74	BGYH Zrt.	0.93
9	FKF Zrt.	0.74	FKF Zrt.	0.93
10	FOKETUSZ Kft.	0.74	FOKETUSZ Kft.	0.93
11	FOTAV Zrt.	0.74	FOTAV Zrt.	0.93
12	FTSZV Kft.	0.74	FTSZV Kft.	0.93
13	BuCC Environment	0.74	BuCC Environment	0.93
14	BuCC Commission	0.65	BuCC Commission	0.59

Index

Page numbers in italics indicate a figure and page numbers in bold indicate a table on the corresponding page. Page numbers followed by 'n' indicate a note.